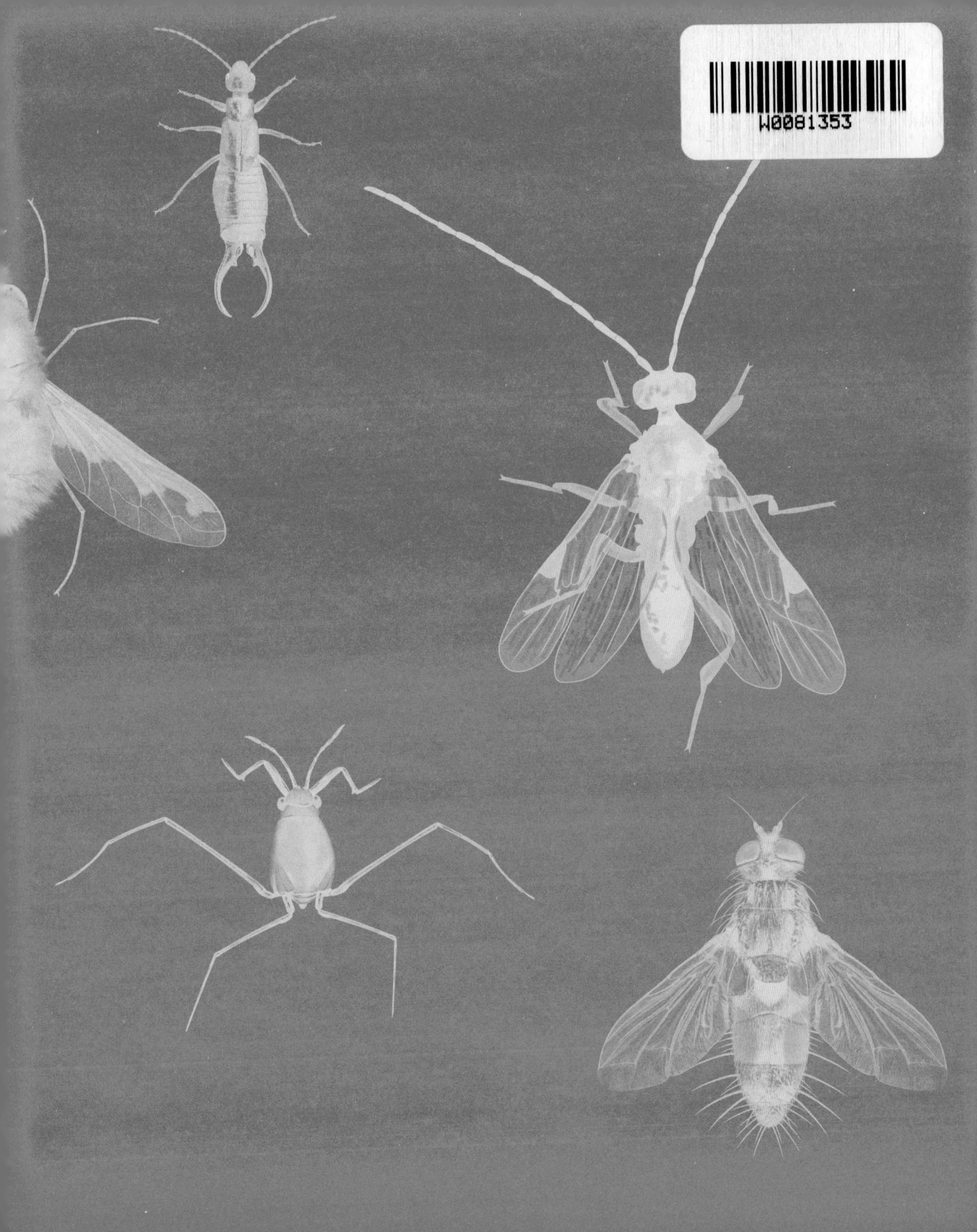

INSECTS

Monarch butterflies
(*Danaus plexippus*)

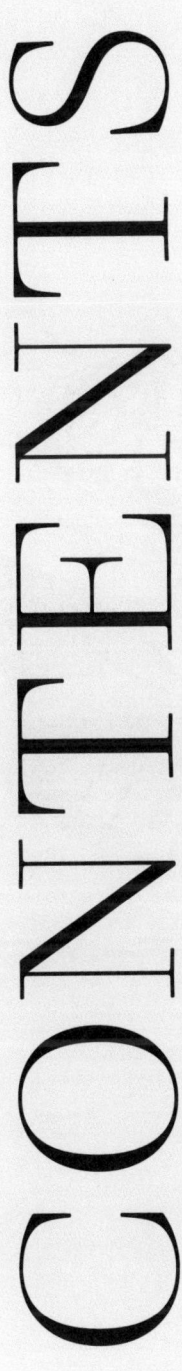

CONTENTS

Insects are the most abundant, diverse, and impactful animals on Earth. There are almost five times more beetle species than all species of birds, mammals, and other vertebrates. So far, about 1 million species have been named, but there are likely another 4.5 million as yet unknown to science, or even more.

WHAT ARE INSECTS?

Insects are small animals with an external skeleton and a body divided into three parts: head, thorax, and abdomen. They mostly have three pairs of jointed legs and two pairs of wings, but there are many exceptions, resulting in an incredible diversity of forms and functions. So what is it about insects that makes them so successful? Six legs have certainly been better than four or two in evolutionary terms. And the ability to fly has clearly been a huge advantage (pp.42–43). In both cases, however, insects show a unique ability to evolve major modifications to their legs and wings. A large proportion of immature insects and some adult insects have no legs, such as adult female scale insects that are sedentary and live in very close association with their host plant. And many adult insects are either wingless or have wings only at certain times, such as when they need to find new hosts or to reproduce.

Possibly the most versatile feature of the insect body is the mouthparts. Made up of several components, insect mouthparts have evolved into a bewildering variety of forms, enabling pollinating insects to reach far inside flowers to extract nectar (pp.86–87), saproxylic insects to easily chew through timber in our houses (pp.38–39), and parasites to quietly suck the blood of larger animals (pp.90–91). At the other end of the body, some insects are able to deliver a sting (pp.94–95). This feature originally evolved in many different types of insects as an ovipositor for laying eggs within host plants, and more than 30,000 wasp species now have the ability to sting their prey.

Insects have many behavioral adaptations, too, most notably in forming societies long before humans walked on Earth. A common feature of social insects (pp.46–47)—ants, termites, and social bees and wasps—is their ability to construct large,

complex nests in which they rear the young. They have also evolved diverse means of communication, including the dance that honeybees perform to pass on information about food sources.

The wide variety of ways in which insects interact with us and with other species is fascinating and inspiring in equal measure. Outside, we regularly encounter insects feeding on plants—although those that live inside plants, such as leaf-mining insects, often go unseen. We notice the pollinators and, of course, the insects that bite and sting us, but we are less aware of the tiny wasps that parasitize other insects and the even tinier wasps that parasitize the parasites. All around us, predatory insects are searching for prey or setting traps for them, and social insects are experiencing some of the most complex ways of life of any animal. Some ants, for example, feed on the honeydew excreted by aphids and mealybugs, often behaving like tiny farmers and protecting these species from predators. In some cases, ants have been seen to move mealybugs onto better plants in the same way a farmer might move their livestock to a better pasture (pp.62–63).

After nearly 500 million years of evolution, insects have adapted to life in almost every habitat on Earth (pp.102–105). The greatest diversity is found in tropical forests, but the skies of temperate regions contain millions of aphids and other insects in search of food. Insects are found in the harshest environments, including scorchingly hot deserts, dark caves, and frozen mountaintops. Many have adapted to life in the soil, and others spend most of their lives underwater. Only the marine environment has remained largely free of insects, but even there, some species— the sea skaters (*Halobates* spp.)—can be found in the middle of the ocean.

Tropical abundance
Forests in the tropics are the most diverse insect habitats on Earth, home to the Rajah Brooke's birdwing butterfly (*Trogonoptera brookiana*, left) and the lanternfly (*Pyrops candelaria*, below).

Nut weevil (*Curculio nucum*)
Acorn weevils (*Curculio glandium*)
and nut weevils live on oak and hazel,
respectively, laying their eggs inside
acorns and nuts, where the larvae
develop. They can be very abundant,
and measures are sometimes needed
to control damage to crops.

From the Ordovician Period onward, insects have evolved to feed on every plant and animal on Earth; to consume even the toughest parts of trees and other plants; and to chew into seemingly indigestible products, such as clothes and carpets. Many of these insects consume a large share of our crops; others, such as mosquitoes and tsetse flies, transmit diseases such as malaria and dengue, which take a huge toll on our health and our livestock. Insect pests, therefore, often have to be controlled by human interventions. Fortuitously, people strive to devise more environmentally sensitive ways of managing pests, using biological and cultural controls, for example (pp.220–221).

Insects affect us in many ways, but it is only a minority that eat our crops and bite us. Through the multitude of interactions they have with other species, insects are critical to the functioning of ecosystems. Without insect pollinators—those that move pollen between plants, fertilizing them and enabling them to produce seeds and fruits—we would not have the diversity of crops we eat today. Without the many natural insect enemies of pest species, such as parasitoid wasps (pp.68–71), and those introduced by farmers for biological control, such as lacewings (pp.74–75), we would face considerably more damage to our food crops. Without insect decomposers recycling the waste in the world's ecosystems, the environment would soon become uninhabitable. And without insects, there would be far less in the natural world to enjoy and explore.

Unfortunately, insects face the same problems as other species on Earth: climate change (pp.196–197), the destruction of their natural habitats, the spread of invasive alien species (pp.198–199), and pollution. Many people are working hard to stop the loss of insect biodiversity by restoring habitats (pp.204–205) or being inspired by the rewilding movement (pp.206–207). Most people will be aware of

efforts to reintroduce mammal species, such as beavers, but entomologists, too, are working to ensure that threatened or even locally extinct insects thrive again. The successful reintroduction of the large blue butterfly (*Phengaris arion*) in England where it became extinct in the 1970s is an inspiring example of how insect declines can be reversed. Saving threatened insects requires finding out why they are in decline, and dedicated research has meant that the large blue butterfly can now be seen again in places where it was lost (pp.208–209).

This book is a captivating exploration of the world of insects: their biology, diversity, and ways of life; their interactions with us and with other species; and the conservation efforts that strive to protect them. But above all, it is a celebration of the fact that insects are essential to life on Earth.

Insects was written by more than 90 members of the Royal Entomological Society, one of the largest societies of entomologists in the world. It has been a joy to work together to write the book and share our enthusiasm for the amazing world of insects. We hope you are as enthralled by the ways of insects as we are, and that some of you—in an amateur or professional capacity—will be inspired to expand our shared knowledge of these "little things that run the world."

Large blue butterfly (*Phengaris arion*)
The large blue butterfly was successfully reintroduced to the UK after research on its close relationship with a species of ant (*Myrmica sabuleti*) provided the knowledge needed for its conservation.

False-leaf katydid (*Eubliastes viridicorpus*)

UNDERSTAND

CHAPTER I

WHAT IS AN INSECT, and how many different types are there? Find out what anatomical features make an insect, inside and out. Discover how these ubiquitous animals came into being and when they evolved into the creatures we recognize today.

Taxonomists have been discovering and naming new species for about 250 years. Of approximately 1.5 million named species—excluding those that have been inadvertently named more than once—the vast majority, or about 1 million species, are insects.

HOW MANY INSECTS ARE THERE?

EARLY ESTIMATE The first estimate of how many insect species there might be globally, including undiscovered species, was made in 1826 by British entomologists William Kirby and William Spence. In the fourth edition of *An Introduction to Entomology*, they suggested there might be 110,000 to 120,000 species of plants and fungi on Earth and, given that they had observed six insect species for each plant, there could be 600,000 insect species worldwide.

RAINING INSECTS In the 1980s, researchers used "fogging" to collect insects from treetops to study. Machines pumping out insecticides were hauled up into the canopy, and the resulting "insect rain" was collected on sheets below. Using this method in Central America, entomologist Terry Erwin suggested there could be 30 million species of insects and other terrestrial arthropods (animals with an exoskeleton, jointed legs, and segmented bodies), including spiders and mites. He thought there were many more species of insects specific to particular tree species, known as host specificity. Nigel Stork and colleagues have since collected data on the level of insect host specificity to trees, as well as on Erwin's other assumptions. They have created mathematical models that suggest much lower numbers of arthropod species on Earth. These, with at least three other methods of calculating global insect diversity, suggest there are 5.5 million insect species and 7 million species of all terrestrial arthropods.

Buff-tailed bumblebee (*Bombus terrestris*)
This is one of around 270 global bumblebee species. As the name suggests, this worker bee sports a buff-colored tail, but as several other species share this trait, individuals can only be accurately identified using DNA testing.

Buffalo treehopper (*Stictocephala bisonia*)
This tiny bug, at just ¼–⅓in (6–8 mm) long, has been named because its head resembles that of a buffalo, complete with black tips that look like horns. It can be found in North America, Europe, Asia, and North Africa.

Sawfly caterpillars (Symphyta)
There are an estimated 8,000 sawfly species worldwide, spanning five insect families. Many of them are specific to one plant species, and they can often be seen feeding from the edge of a leaf, hanging on with their front legs.

*The diversity of insects is remarkable and their evolution
no less so. The latter has led to an astounding range of unique
and incredible life histories, illustrating myriad ways
that insects have perfected life on Earth.*

INSECT EVOLUTION

GRAND HISTORY The staggering number of insect species is widely known, but less well known is the fact that insects are also of great antiquity, and their history comprises over 400 million years of innovation. They were, for example, the first to fly, first to sing, first to be used within agriculture and animal husbandry, and first to achieve abstract communication. In fact, insects were among the earliest terrestrial animals and therefore experienced the mass extinction events that humbled trilobites (marine arthropods), ammonites, and nonavian dinosaurs. But climate change, invasive alien species, habitat fragmentation and destruction, pollution, and overexploitation may yet prove too much.

TAKING TO THE SKIES

At some point, relatives of silverfish evolved a means of gliding between plants using extensions of the thorax, and over time these became articulated to form wings. Today, more than 95 percent of all insects are capable of taking to the air. But wings have evolved into more than a means of movement, with forms that can be used for thermoregulation, camouflage, communication (attracting mates and warning predators), or shielding the body.

EARLY INSECTS The earliest insects were small, flightless animals that fed on the nutritious spore-forming structures of early plants or possibly scavenged for detritus. The modern survivors of these early insects are today's bristletails and silverfish (pp.144–145). Some 60 to 70 million years after their origins, insects evolved complete metamorphosis: in which the life cycle is compartmentalized into egg, larva, pupa, and adult (pp.28–29). This effectively allowed the stages to live unique lives in different habitats and on different diets. Such insects define insect diversity today, and include the truly hyperdiverse lineages: wasps, bees, and ants (pp.164–165); beetles and weevils (pp.170–171); flies, mosquitoes, and gnats (pp.174–175); and moths and butterflies (pp.176–177).

EVOLUTIONARY HISTORY OF INSECTS

1.

Insects originated approximately 410 million years ago, a long time after terrestrial plants.

2.

They share a common ancestor with Entognatha (a class of wingless arthropods). The ancestor of both emerged onto land after evolving from marine crustaceans.

ROCK BRISTLETAIL
(*MACHILOIDES*)

LARGE BROOK DUN
(*ECDYONURUS TORRENTIS*)

3.

The earliest evidence of flight is the fossil *Rhyniognatha hirsti*, a winged insect from 410 million years ago. Around the same time, a group of insects diverged from silverfish and took flight.

4.

Winged insects underwent a major radiation (evolving in different directions and adapting to different habitats) some 356 to 299 million years ago, leading them to diversify into many different forms.

5.

A major radiation of insects that underwent complete metamorphosis followed, occurring from 299 to 252 million years ago.

6.

About 252 million years ago, a mass extinction event wiped out many early insects, with those surviving being today's major insect orders.

7.

Several insect orders, such as Lepidoptera (butterflies and moths) and Coleoptera (beetles), diversified with flowering plants.

*Tailored by evolution for life in almost every habitat
on Earth, the insect body is among the most
adaptable structures in nature.*

THE INSECT
BODY PLAN

Insects comprise three main body segments: head, thorax, and abdomen. The head houses the compound eyes, brain, antennae, and external mouthparts. The thorax holds the legs and wings (pp.18–19), and the abdomen holds many internal organs, including the gut and reproductive system (pp.20–21). Insects are hexapods, meaning they have six legs. These are attached to the underside of the thorax. At the top of the thorax are crisp membranous wings, derived from extensions of the cuticle (see below). The wings are powered by muscles in the thorax, which can contract up to several thousand times per second during flight.

EXOSKELETON The insect body is contained in a tough exoskeleton made mostly of a protein called chitin. This shiny armor is commonly referred to as the cuticle, and it supports and protects the insect. The cuticle expands as the insect grows, but because the cuticle can only withstand a limited amount of stretching, the insect undergoes a process called molting. At each molt, the insect sheds the remains of its old cuticle—an act known as ecdysis—and replaces it with a new one.

INTERNAL STRUCTURES Inside the exoskeleton, important structures, such as breathing tubes (trachea), muscles, and bundles of nerves (ganglia), may be repeated within each body segment. Insects have an entirely different circulatory system to humans and other vertebrates (pp.20–21), and to allow them to breath oxygen throughout their bodies, trachea are needed in each segment. Control of the insect body relies on communication between the brain

and the ganglia in each body segment (pp.20–21). The arrangement of signals within the ganglia is important for cyclical functions, such as walking and flying, in which structures have to move repeatedly in a special pattern.

REPEATING UNITS Insect diversity is partly down to their body being formed of many repeating units, allowing for similar structures to evolve multiple times in different parts of the body. Legs in adult insects, for example, usually occur in pairs: two legs per segment of the thorax. Insect larvae also have thoracic legs, but in addition, they have pairs of legs (usually called prolegs) on one or more of their abdominal segments. Most caterpillars, for example, have five sets of abdominal legs, including a set at the rear known as anal prolegs. However, caterpillars in the moth family Geometridae only have two sets of legs: the anal prolegs and another pair close to the rear. This means they have a different way of walking, forming loops as they move forward, giving rise to several common names such as loopers, inchworms, and measuring worms.

HEAD   THORAX   ABDOMEN

Insect segments
All insects have the same body plan, and the three parts are most clearly seen in adult insects, such as this European beewolf (*Philanthus triangulum*).

The head is the first of three major segments of the insect body. It serves as the "command center" and houses antennae, eyes, and a variety of mouthparts.

THE INSECT HEAD

The head is attached to the thorax by the neck (cervix), which is membranous and receives muscular support, thus enabling the head to move. Externally, the rigid capsule protecting the head is made up of multiple plates of cuticle comprising proteins and chitin, supported internally by a framework of struts (tentoria). From top to bottom, the frontal plates are named the vertex, the frons, and the clypeus. In the more primitive forms of insects, the head orients downward, with the vertex at the top and the mouthparts lowermost. This is common in plant-eating insects, such as locusts. In predatory insects or those that feed by chewing, such as ladybugs, the mouthparts are oriented forward.

MOUTHPARTS Mouthparts differ depending on how the insect feeds and the type of food it eats. They have adapted to the requirements of chewing, sucking, or lapping, for example. The labrum (essentially an upper lip) protects the other segments of the mouthparts. These include the mandibles, which cut and chew food; the maxillae, which act like pincers to hold food; and two structures called the galea and the lacinia, which are involved in chewing. Some insects have a long feeding tube, known as a proboscis, which is formed by the elongated maxillae in many species. Examples include moths and butterflies, which coil up their proboscis when they are not feeding.

ANTENNAE The antennae are highly specialized sensory organs that vary greatly among insect orders. Primitively, the antennae comprise three segments. The basal segment (scape) inserts into the socket, the second segment is the pedicel, and the third is the multisegmented flagellum. Usually, it is the flagellum that varies most in length and shape. There are many thoughts on why antennae

are so diverse, including that the surface area and shape can alter the flow of air across the antennae and therefore influence the ways in which insects detect chemicals. More simply, larger antennae have more space for more sensory cells.

EYES Most adult insects have two large eyes, one on each side of the head. These eyes are compound, meaning they are made up of separate units, each with a lens and a few sense cells. The number of units (ommatidia) varies, and dragonflies have more than 10,000, for example. The eyes tend to bulge outward, allowing the insect a wide field of vision, with binocular vision in front, above, and below. Adult insects usually also have three ocelli, very simple eyes that detect movement and light intensity but do not produce an image. Insect larvae do not have compound eyes, but have two to eight single-lens eyes called stemmata, which evolved from compound eyes.

Head anatomy
The heads of adult insects vary greatly, depending on taxonomy (insect order or family), habitat, and mode of feeding.

Ground beetle (*Notiophilus biguttatus*)
Ground beetle mouthparts are made up of a large labrum in the center, with the two thin parts of the labium visible immediately below. There are pairs of maxillae and mandibles on either side.

Dark spreadwing (*Lestes macrostigma*)
Damselflies have very large compound eyes on the side of their head. Their antennae are relatively short, so they rely on vision to navigate and find prey.

Large heath butterfly (*Coenonympha tullia*)
Butterflies use their antennae to sense certain chemicals in their surroundings and seek out potential mates and sources of food. Antennae also help with balance and detecting motion.

The basic insect body plan is similar for all insects,
but there is huge variation in the morphology
of many of their external structures.

EXTERNAL ANATOMY: THORAX AND ABDOMEN

WINGS There are wingless insect species, but most insects have four wings, modified in various ways. In some species, both pairs of wings are similar in size and shape, but in most, one of the wing pairs has become modified. For example, the forewings of many insects, such as beetles, have a protective function, and the hind wings of others can be very small; in the true flies, they are reduced to small organs called halteres. The wings of numerous insects, such as dragonflies and wasps, are semitransparent membranes, whereas those of moths and butterflies are covered in tiny, overlapping scales. Although most insects use wings for flight, many also use them to make sound.

LEGS With very few exceptions, adult insects have six legs—three pairs attached to each segment of the thorax. A typical leg has six parts: coxa, trochanter, femur (the largest and stoutest part in most insects), tibia, tarsus (split into several tarsomeres), and pretarsus (with a claw or claws). The basic walking leg may be modified according to various functions, including jumping, swimming, digging, and making noise. Predators' legs are modified in different ways for grasping prey. For example, the femur and tibia of mantid legs are formed into pincers, with spines along both parts. Ectoparasitic insects, such as lice, also have adaptations, usually well-developed claws for grasping the hairs of their hosts.

OVIPOSITOR This hollow structure is used for egg laying, and it shows incredible morphological variation. For example, some grasshoppers have a thick, sturdy ovipositor that looks and functions like a plow, enabling the insect to lay eggs into the ground. In contrast, parasitoid wasps have a long, needlelike ovipositor, which they use to pierce the bodies of other (often much larger) insects to lay their eggs inside. The ovipositors of bees, wasps, and ants have evolved to enable these insects to sting other animals by injecting venom.

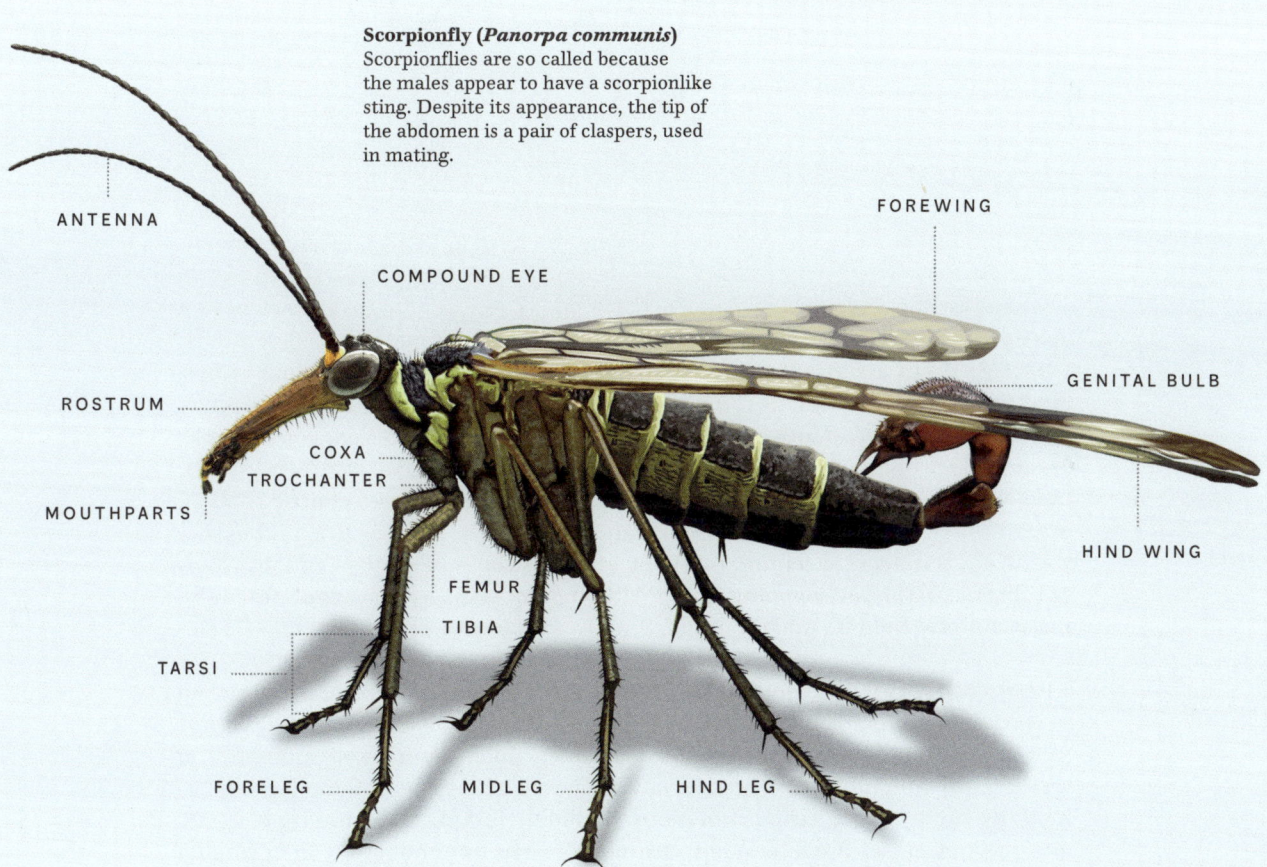

Scorpionfly (*Panorpa communis*)
Scorpionflies are so called because the males appear to have a scorpionlike sting. Despite its appearance, the tip of the abdomen is a pair of claspers, used in mating.

ANTENNA

FOREWING

COMPOUND EYE

ROSTRUM

GENITAL BULB

COXA
TROCHANTER

MOUTHPARTS

HIND WING

FEMUR

TIBIA

TARSI

FORELEG

MIDLEG

HIND LEG

The internal anatomy of insects is very different from that of vertebrates because they do not deliver oxygen via a circulatory system or pump blood around the body in a closed vessel system.

INTERNAL ANATOMY

RESPIRATORY SYSTEM Unlike vertebrates, which have a distinct gill or lung structure, an insect's respiratory system comprises a network of tubes (trachea) throughout the body that transport oxygen to each segment individually. They connect to the outside through tiny holes called spiracles, which allow for the exchange of oxygen and carbon dioxide. Small insects rely almost entirely on diffusion for moving gases within the tracheal system, but larger insects have a more complex arrangement. By opening and closing some spiracles sequentially, they can actively transport gases. In all cases, limitations to the flow of gases determines the insect's size, because only so much oxygen can be taken in through the trachea. In the Carboniferous, some 300 million years ago, when oxygen concentrations were higher than they are today, insects were much larger. Dragonflylike insects, for example, had wingspans of more than 24 in (60 cm).

OPEN CIRCULATION Instead of blood, insects have a transparent, nutrient-rich liquid called hemolymph. It flows freely through cavities within the body, making direct contact with internal tissues and organs. This open circulatory system is in contrast to the closed system of humans and other vertebrates, in which blood is pumped by the heart through closed vessels. Insects have an equivalent structure to the heart, which pumps hemolymph through a muscular tube called an aorta.

A DIFFERENT KIND OF MIND Insects do not concentrate their entire nervous system in their brains, but instead have clumps of nerves called ganglia throughout the body. The arrangement of signals within the ganglia is important for cyclical functions, such as walking and flying, in which structures have to repeatedly move in a specific pattern. The number of ganglia varies: in fruit flies, they fuse into a single clump, while in honeybees, they remain mostly separated

into seven distinct bundles of neurons. Each ganglion has striations that extend like tentacles into the flesh of the insect. These control hubs can act independently of the brain. In fact, decapitated grasshoppers can live and move around for some time without a head.

DIGESTION The insect alimentary canal is made up of the foregut, midgut, and hindgut. Comprising several functional parts, including the crop, the foregut is used to store food and sometimes to fragment it. The production of digestive enzymes, digestion, and the uptake of digestive products take place in the midgut. Undigested food passes through the hindgut and is excreted via the anus. There are various adaptations to the alimentary canal to cope with the diverse diets of insects. Plant-sucking insects (such as aphids) and blood-sucking insects, for example, consume large quantities of water and have anatomical and behavioral adaptations to ensure that their hemolymph does not become too dilute.

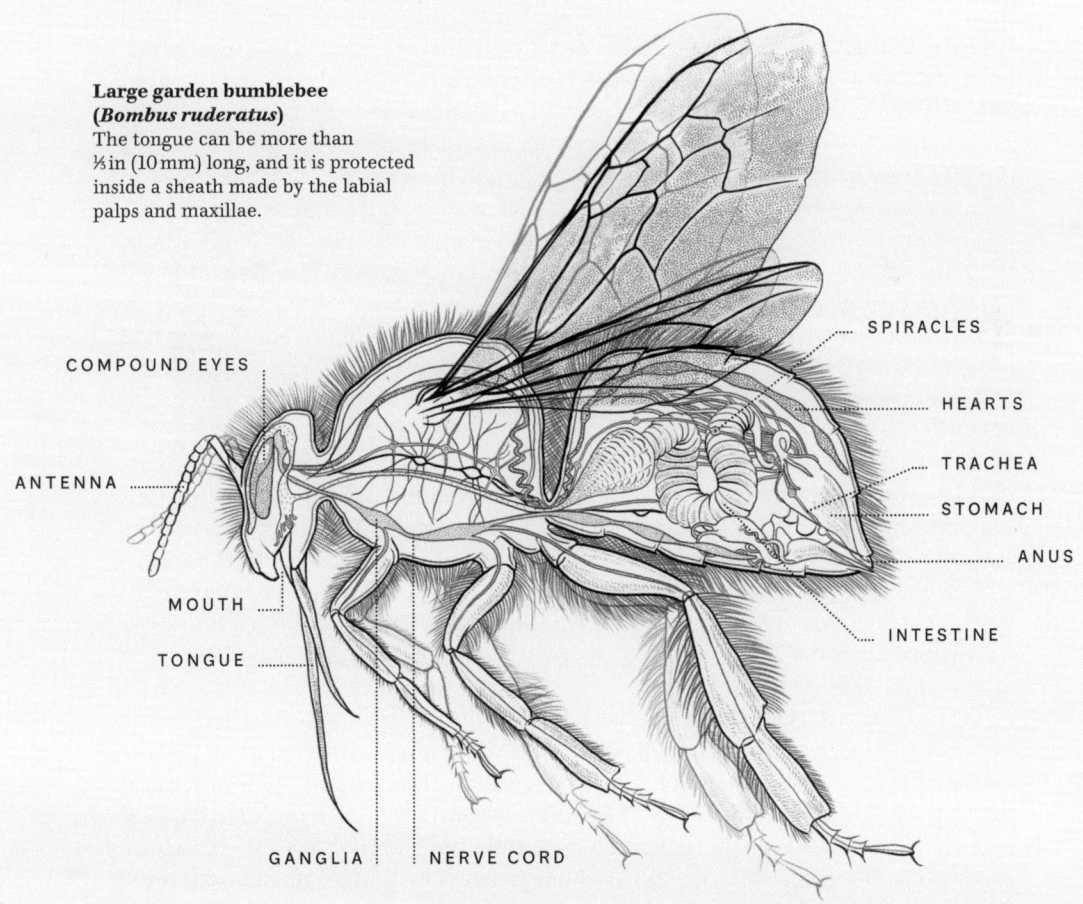

Large garden bumblebee (*Bombus ruderatus*)
The tongue can be more than ⅖ in (10 mm) long, and it is protected inside a sheath made by the labial palps and maxillae.

COMPOUND EYES

SPIRACLES

HEARTS

ANTENNA

TRACHEA

STOMACH

ANUS

MOUTH

INTESTINE

TONGUE

GANGLIA NERVE CORD

**False-leaf katydid
(*Eubliastes viridicorpus*)**
A newly discovered species of false-leaf katydid has large ears on the end of its very long legs. Their position is useful because it increases the distance between the ears, and thus gives the insect accurate directional hearing.

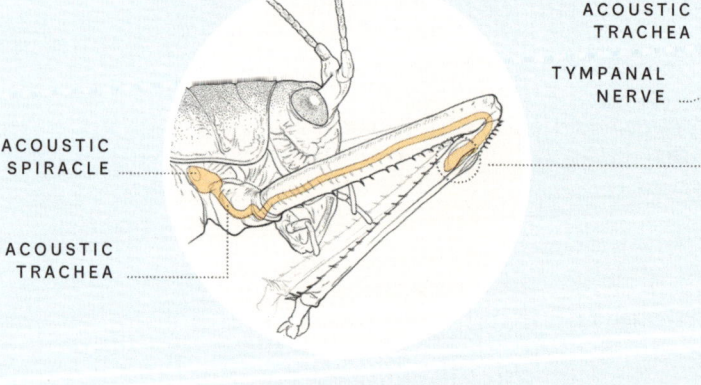

ACOUSTIC
TRACHEA

TYMPANAL
NERVE

ACOUSTIC
SPIRACLE

ACOUSTIC
TRACHEA

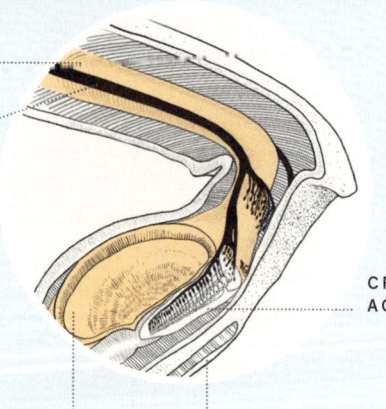

CRISTA
ACUSTICA

TYMPANUM   PINNA

Tympanal ear
As sound reaches the eardrum, it vibrates a liquid-filled organ called the crista acustica, which functions just like the human cochlea. High-pitched sounds are received at one end of this organ, and low-pitched sounds at the other—like a tiny piano. This mechanism is known as tonotopy, and it allows the bush-cricket to discriminate between different sounds.

Ear canal
Bush-crickets have ear canals like humans. This ear canal is known as the acoustic trachea, and it delivers sound to the eardrums from an opening in the thorax called the acoustic spiracle.

The first insects are thought to have been deaf, but insect ears subsequently evolved more than 20 times. They are found on almost every part of the insect body and are used to find mates, detect predators, and size up rivals.

HEARING

There are two types of insect ear, and they each function in very different ways. In some insects, such as fruit flies and honeybees, the antennae serve as sound receptors. A structure called the Johnston's organ, located at the base of the antennae, detects tiny differences in how air particles move in response to sound. This organ bends in various ways depending on the direction the sound is coming from. The more common type of insect ear is the tympanal ear, comprising a membrane stretched across a frame—the tympanal organ. Tympanal organs can be found in different parts of the insect body depending on the species, including at the base of the wings or legs. Insects with tympanal organs detect differences in sound pressure as the membrane vibrates. Both types of insect ear involve special sensory cells called mechanosensors, which convert physical stimuli into nerve signals.

EARS LIKE OUR OWN One of the most remarkable insect hearing systems belongs to the bush-cricket. Like humans, bush-crickets hear with eardrums (tympana). However, as with many insects of the Orthoptera order (pp.150–151), bush-cricket eardrums are located in the front legs—two in each leg. These tympanal ears are used primarily to detect species-specific songs of male bush-crickets, which make sounds using their wings. They are also used for detecting predators, and some species have external structures called pinnae for bat detection.

Nest of African paper wasp (*Belonogaster juncea*)

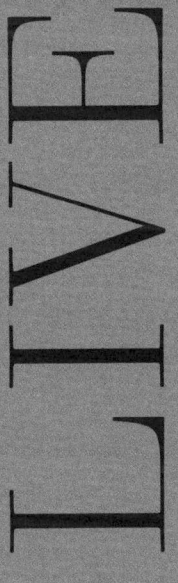

CHAPTER II

LIVE

INSECTS CARRY OUT A MULTITUDE OF FUNCTIONS in the natural world. Their lives are amazing and full of surprises. Discover the aphids that carry two generations of young inside them, the migratory species that follow the paths of their ancestors to unknown places, the fireflies that glow in the dark, and the intricate insect cities that are built beneath our feet.

Most insects undergo a metamorphosis and transform from one body shape to another. Entomologists categorize insects according to how they develop, the vast majority falling into one of three categories: ametabolous, hemimetabolous, and holometabolous.

LIFE CYCLES

AMETABOLOUS These insects do not metamorphose as part of their life cycle, and there is little change in the external anatomy of the adult compared to earlier stages. Jumping bristletails and silverfish, for example, emerge from the egg looking like miniature adults. Important internal changes occur to allow reproduction, and ametabolous insects are only able to reproduce after undergoing several molts (shedding their exoskeleton). The adults then continue to molt, and successive adult stages go on reproducing.

HEMIMETABOLOUS True bugs and grasshoppers are examples of hemimetabolous insects. They emerge from the egg as immature nymphs, which look similar to the adult but do not have wings. Such

Hemimetabolous development
The kissing bug (*Rhodnius prolixus*) goes through five nymph instars (stages) before becoming an adult.

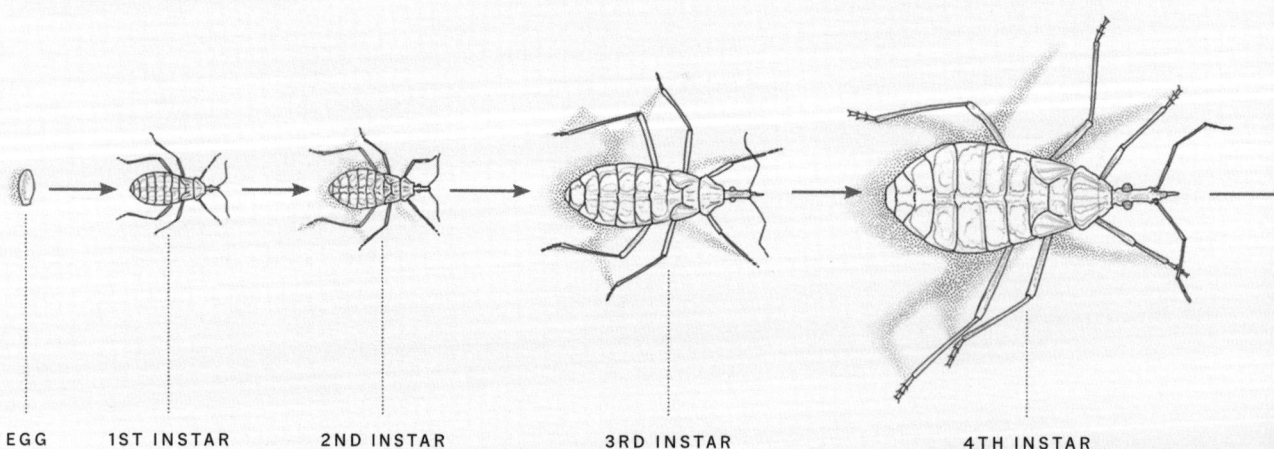

EGG   1ST INSTAR   2ND INSTAR   3RD INSTAR   4TH INSTAR

insects are said to undergo a simple or incomplete metamorphosis
(pp.30–31). Their wings develop within external wing buds, and
hemimetabolous insects do not molt as adults. Mayflies, which molt
once from the preadult stage to the mature adult form, are an
exception. The mayfly preadult, which transforms from the nymph at
the water surface, may be regarded as specialized to allow the adult to
emerge successfully from the water.

HOLOMETABOLOUS These insects undergo a complete
metamorphosis, hatching from the egg to reveal a soft-bodied larva
(pp.28–29). Examples include beetles, butterflies and moths, flies,
wasps, bees, and ants. Because the holometabolous larva is so
completely different from the adult, a special intermediate stage
(pupa or chrysalis) is formed to facilitate transformation to the adult
stage. Complete metamorphosis allows the larva to adopt a different
lifestyle to that of the adult, eating different food in a different
environment. Its simplified anatomy enables rapid growth, which
reduces the risks from predators and parasites.

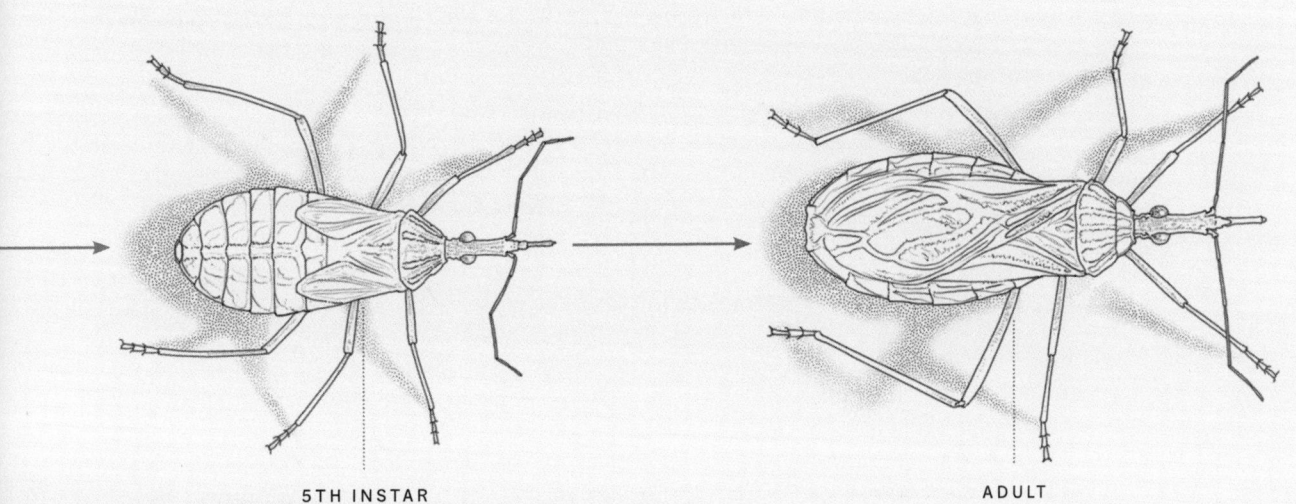

5TH INSTAR   ADULT

*Complete metamorphosis refers to the
development of insects through four separate and distinct
life stages: egg, larva, pupa, and adult.*

COMPLETE METAMORPHOSIS

Insects that undergo complete metamorphosis are termed "holometabolous," and they include beetles, butterflies and moths, flies, and fleas, as well as wasps, bees, and ants. A key distinguishing feature is their soft-bodied larvae, which are dramatically different in appearance from the adult insect. Also notable is the pupal stage, which comes between larval and adult stages. It is during the nonfeeding pupal stage that insects undergo their metamorphosis.

STAG BEETLE The European stag beetle (*Lucanus cervus*) is an excellent example of an insect that develops through complete metamorphosis—but it takes a very long time. Larvae emerge from an egg stage after about three weeks and live entirely within decaying wood, especially oak. They have a hardened head capsule (helmetlike covering) and legs, but other body parts are soft and vulnerable. Their simple shape is well adapted to burrowing, and after three or four years, they leave the wood to build an underground pupation chamber. After shedding their pupal cuticle (tough outer covering) in the fall, adults then overwinter in the soil before emerging in the spring. Despite their huge size (they are the UK's largest insects), stag beetles are strong flyers. Females disperse to find suitable dead wood in which to lay eggs, while males search for females to mate. Males use their spectacularly enlarged jaws to display to females and to fight other males. Both sexes engage in defensive displays and can bite, but they are not dangerous to humans. Adult stag beetles live for only a few weeks, and both males and females die soon after the eggs are laid.

LIFE CYCLE OF THE STAG BEETLE

Egg
About ⅛ in (3 mm) long, eggs are laid individually or in small groups in compacted soil, close to decaying wood.

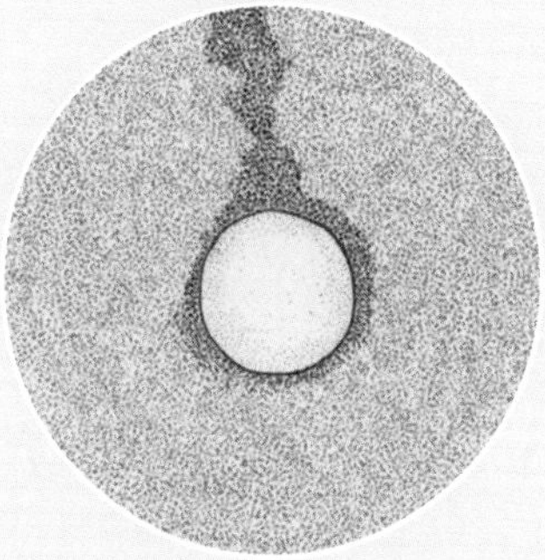

Adult
The adult stag beetle leaves the pupa in fall, then overwinters underground before emerging the next spring to find a mate. The cycle then begins again.

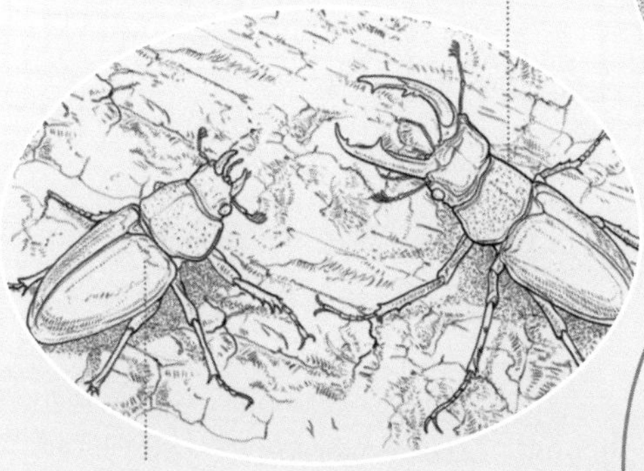

MALE

FEMALE

First instar larva
At this instar (developmental stage), the larva is still about ⅛ in (3 mm) long. It feeds for several weeks before molting (shedding its exoskeleton).

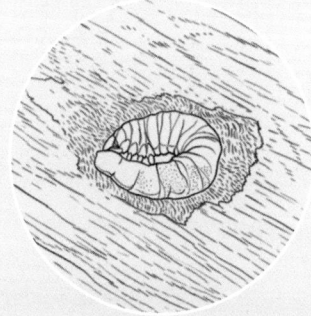

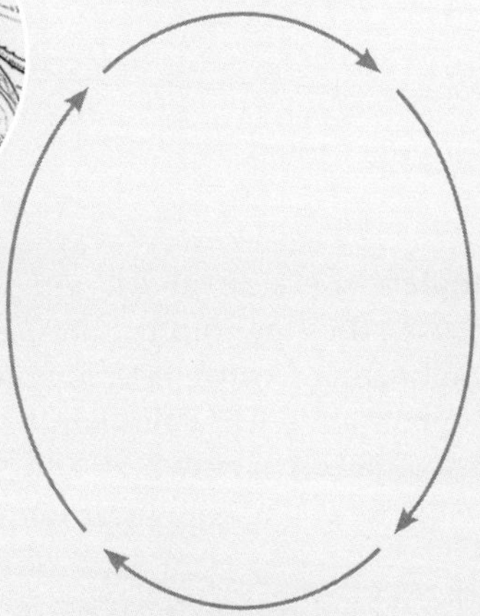

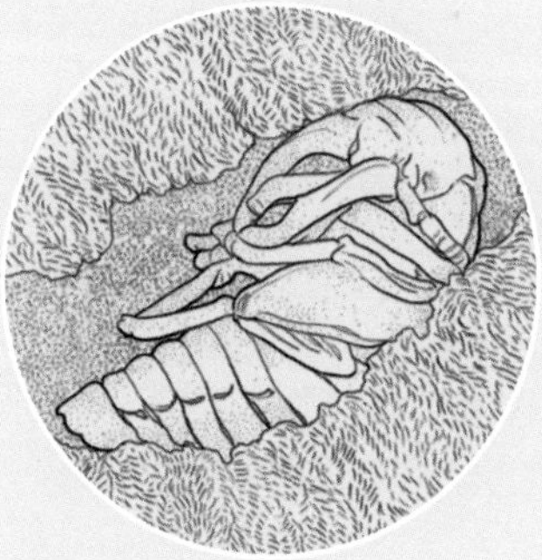

Pupa
When growth is complete, the larva burrows into nearby soil to pupate. The pupa is significantly smaller than the larva.

Second instar larva
It takes several months for the larva to grow to 1½ or 2 in (4 or 5 cm) in length. Growth is usually complete by the end of the first year.

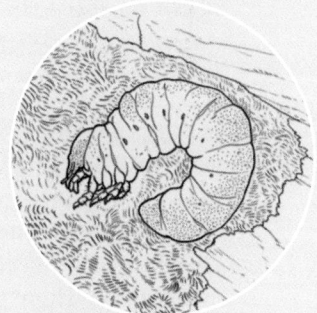

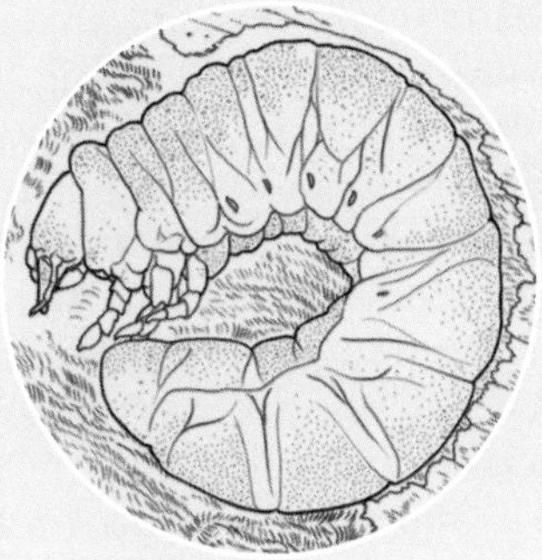

Third instar larva
This stage lasts two or three years. The mature larva may be 3 in (8 cm) long and weigh more than ¾ oz (20 g).

The second main type of development for young to reach adulthood is incomplete metamorphosis. This refers to the development of insects through three life stages: egg, nymph, and adult.

INCOMPLETE METAMORPHOSIS

Insects that develop through incomplete metamorphosis are termed "hemimetabolous," and they include all true bugs, crickets, grasshoppers, and cockroaches. The key distinguishing feature of incomplete metamorphosis is that, after emerging from its egg, the nymph resembles a miniature adult.

HAWTHORN SHIELD BUG The hawthorn shield bug (*Acanthosoma haemorrhoidale*), which is widespread throughout Europe, is an example of an insect that develops through incomplete metamorphosis. As the name suggests, this species feeds mainly on hawthorn—both the berries and the leaves—although sometimes on other trees, too, such as oak, birch, holly, and hazel. Both sexes overwinter as adults and emerge from winter dormancy to mate in the spring. Mated females then lay clusters of eggs on the underside of host leaves, and the newly hatched nymphs appear a few days later. These nymphs look like adult insects but lack wings and the capacity to reproduce. They develop gradually through five nymphal instars (stages), which can be identified based upon size and patterning. Between each instar, the insect molts—sheds its cuticle (tough outer covering)—which allows it to grow. The nymphs can walk to new feeding sites, but only once the wings have developed after the fifth instar and final molt does the hawthorn shield bug become an adult. It can then fly to find mates and new places to feed.

LIFE CYCLE OF A HAWTHORN SHIELD BUG

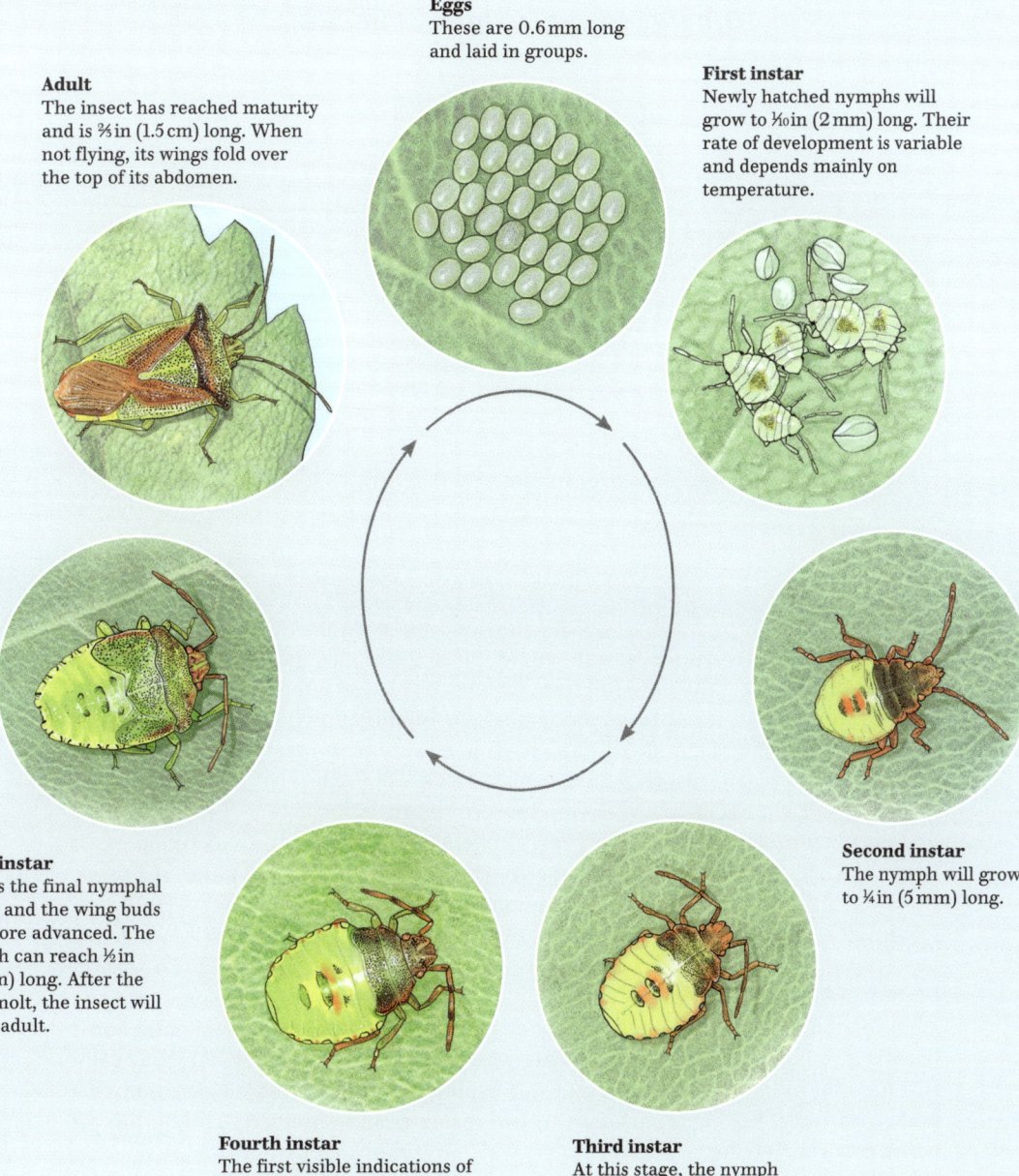

Eggs
These are 0.6 mm long
and laid in groups.

First instar
Newly hatched nymphs will
grow to ¹⁄₁₀ in (2 mm) long. Their
rate of development is variable
and depends mainly on
temperature.

Adult
The insect has reached maturity
and is ⅗ in (1.5 cm) long. When
not flying, its wings fold over
the top of its abdomen.

Second instar
The nymph will grow
to ¼ in (5 mm) long.

Fifth instar
This is the final nymphal
stage, and the wing buds
are more advanced. The
nymph can reach ½ in
(1.2 cm) long. After the
next molt, the insect will
be an adult.

Fourth instar
The first visible indications of
the wings—wing buds—appear at
this stage, and the nymph can
grow to ⅖ in (1 cm) long.

Third instar
At this stage, the nymph
reaches ⅓ in (8 mm) long.

Part of the success of aphids is because of their unique and unusual life cycle. Aphids have very short generation times and fast maturation, enabling populations to build up rapidly when environmental conditions are favorable.

APHIDS: MOTHERS AND DAUGHTERS

BIRTHING GRANDDAUGHTERS The females of many aphid species are capable of producing offspring without the need to mate with a male. These females reproduce asexually and give birth to genetically identical clones of themselves. Furthermore, these daughters are born with their daughters already developing inside, in a phenomenon known as the "telescoping of generations," thus making aphids the Matryoshka dolls of the insect world.

HOST PLANTS Some aphid species remain on one host plant throughout the year, but others (about 10 percent) spend the summer on one plant and the winter on another. Usually, those that alternate spend the summer on a herbaceous plant and the winter on a tree. The bird cherry-oat aphid (*Rhopalosiphum padi*), for example, spends the summer on grasses or cereal crops and the winter on bird cherry trees. Because the quality of tree leaves (as food) drops in the middle of summer, switching hosts avoids this issue. Some aphid species that remain on their tree hosts, such as the sycamore aphid (*Drepanosiphum platanoidis*), cope by going into a state of suspended animation (aestivation), the summer equivalent of hibernation. Alternating between host plants may also have evolved to allow aphids to escape from their many natural enemies, such as parasitoid wasps (pp.68–69), which regularly attack them.

APHID DEFENSES

Aphids have several strategies to defend themselves against their natural enemies. They produce an "alarm" chemical to warn other aphids if they are being attacked, and they also have special organs (siphunculi) that can produce a quick-hardening wax to gum up the jaws of their attackers, such as ladybugs.

TO FLY OR NOT TO FLY Unlike most adult insects, many aphid species, including crop pests such as the English grain aphid (*Sitobion avenae*) and the rose-grain aphid (*Metopolophium dirhodum*), have individuals without wings. As plants become crowded with aphids, these species give birth to winged individuals. Not only are these aphids equipped with wings to take them to new plants, but they also have longer antennae and other features to help them get there. Aphids tend not to bother with wings in the winter when laying eggs (in those species that oviposit), and again in the spring when they focus on producing as many offspring as possible.

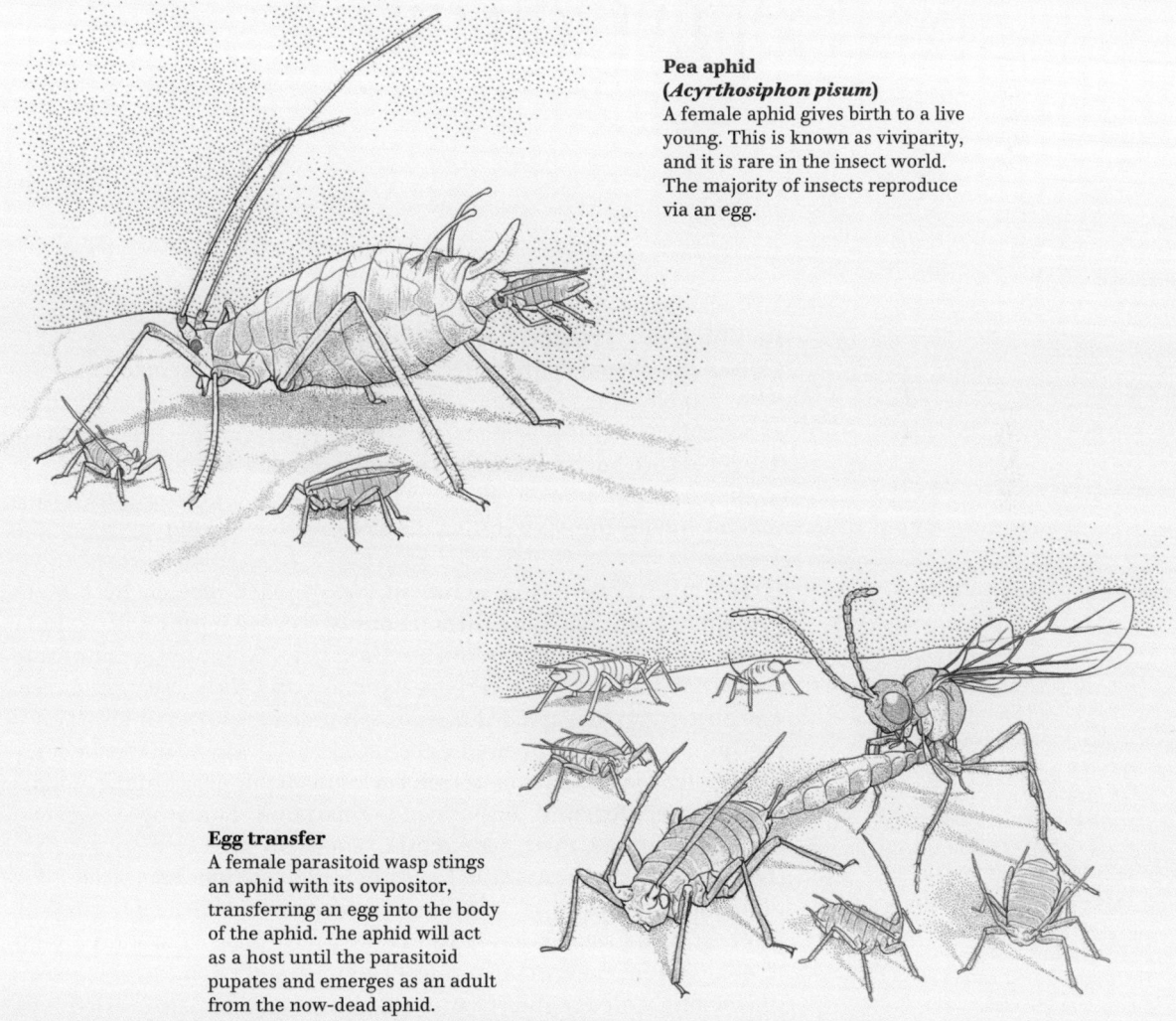

Pea aphid
(***Acyrthosiphon pisum***)
A female aphid gives birth to a live young. This is known as viviparity, and it is rare in the insect world. The majority of insects reproduce via an egg.

Egg transfer
A female parasitoid wasp stings an aphid with its ovipositor, transferring an egg into the body of the aphid. The aphid will act as a host until the parasitoid pupates and emerges as an adult from the now-dead aphid.

In the natural world, mimicry is a fascinating way for insects to reduce—but not eliminate—predation. It occurs when one prey species evolves to resemble another, usually with one or both being "noxious" (detrimental) to their predators.

MIMICRY

Traditionally, mimicry is divided into two main types, called Batesian and Müllerian after their discoverers. Batesian mimicry is where a harmless prey (the mimic) evolves to resemble a noxious prey (the model), thereby gaining protection from predators that have learned to avoid such insects. Good examples are the various species of bee hawkmoth. Müllerian mimicry is where a set of noxious species evolve to resemble one another in order to share the burden of teaching avoidance to their predators.

BATESIAN MIMICRY The success of Batesian mimics depends on how common the model is relative to the mimic. It also depends on how closely the mimic resembles the model, because the predator needs to make mistakes for the "system" to work. The noxiousness of the model is important, too: the more horrible the experience, the more careful the predator will be to avoid insects that even vaguely resemble the model. This is why there are many more mimics of wasps than of bumblebees. A small bird such as a robin may not be killed by a wasp sting, but it will feel too unwell to feed for a day or two, which may make the difference between life and death.

MÜLLERIAN MIMICRY These mimics evolve to form a small number of mimicry "rings." A ring is a group of species that resemble one another, sharing a warning signal—such as a color pattern—to deter predators. In one environment, there will be several such rings. For example, there will be three to four bumblebee mimicry rings in any one habitat. In the UK, there are three common warning patterns: black with a red "tail"; black, white, and yellow bands; and all tawny.

Hoverfly (*Volucella bombylans* var. *bombylans*)
This large species of hoverfly has two main morphs (varieties). This one mimics the red-tailed bumblebee (*Bombus lapidarius*).

Hoverfly (*Volucella bombylans* var. *plumata*)
This morph mimics another bumblebee species, the buff-tailed bumblebee (*Bombus terrestris*).

Light organ
The yellowish-green light organ is at the tip
of the firefly's abdomen.

Fireflies in a forest in Tennessee
Synchronous fireflies flash in a distinct 5–8 signal, synchronizing
when they are dark, not flashing.

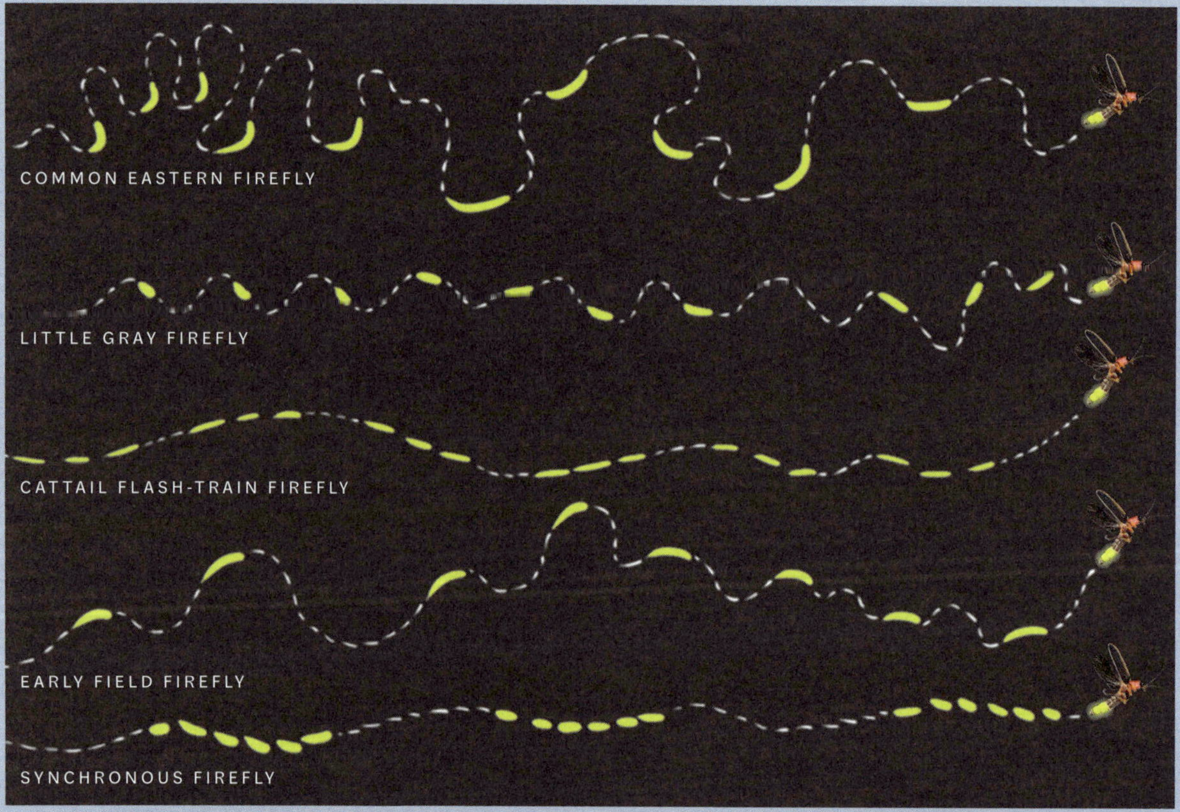

COMMON EASTERN FIREFLY

LITTLE GRAY FIREFLY

CATTAIL FLASH-TRAIN FIREFLY

EARLY FIELD FIREFLY

SYNCHRONOUS FIREFLY

Distinctive flashes
Different patterns of light are emitted by different species of *Photinus* firefly.

*Some families of insects—mainly beetles—have the
ability to emit light and glow in the dark using
a process called bioluminescence.*

GLOWING IN THE DARK

HOW DO INSECTS EMIT LIGHT? All insects make light in the same way, via a chemical reaction. A specialized light-emitting organ contains an enzyme called luciferase, which breaks down the chemical luciferin when in the presence of oxygen, causing a "cold" visible light to be emitted. Periodically restricting the air supply produces flashes of light. Most bioluminescent insects produce light at wavelengths that appear green to humans, but some have more than one luciferase, so they emit light of different colors.

WHY DO INSECTS EMIT LIGHT? Bioluminescent beetle larvae glow continuously. Their light warns would-be predators that the insect contains distasteful or toxic defensive chemicals. Many adult luminescent beetles also use light as a sexual signal. In these species, females are wingless and must attract winged males. Fireflies of both sexes produce species-specific patterns of flashes. Females mate only with males that flash in the correct pattern.

The fireflies of the genus *Photuris* are among a number of beetles that use light as a lure to capture insect prey. While most bioluminescent insects are specialized predators of either insects or snails, only *Photuris* fireflies use light to mimic the flashing pattern of other firefly species to attract males, which they then catch and eat. The predatory female not only gains a valuable food source, but is also able to steal the anti-predator toxins from the bodies of the captured males. This, in turn, protects her against spiders and other predators.

Vast communities of insects are silently working to remove decaying organic matter from the environment. Without this team of recyclers, our world would look and smell very different.

DECOMPOSERS

The remains of dead plants and animals are nutritiously rich sources of food for many insects. This group of rot lovers are collectively known as saprophages, from the Greek *sapros* meaning "rotten" and *phagein* meaning "to eat or devour."

DEADWOOD DWELLERS A complex community of insects, fungi, and microorganisms contribute to the decomposition of wood. In fact, about 10 percent of all species are saproxylic, meaning that they depend on decaying wood for at least some part of their life cycle (pp.110–111). In the UK, for example, there are approximately 650 saproxylic beetle species. The tunneling and feeding activity of wood-boring beetles speeds up decay and attracts a succession of additional beetles and other insects that prefer increasingly rotten wood.

FEASTING ON FECES Globally, there are more than 7,000 species of dung beetle: from poop pushers that roll dung balls and can bury 250 times their own bodyweight, to tiny (<⅓in/10 mm) dung dwellers that never leave the pat. By actively removing dung, these beetles provide a range of essential ecosystem services. They enhance the quality and structure of the soil, improve drainage, reduce greenhouse gases, and lower parasite burdens in livestock. Additionally, they themselves are an important food source for other wildlife.

CARRION FEEDERS A variety of insects are attracted to carrion (animal remains). During the initial stages of decay, blow flies, flesh flies, and carrion beetles will colonize the carcass, their larvae feasting on the soft flesh. This draws in predatory species, such as rove beetles and clown beetles, which benefit from the writhing mass of insect life. Nothing is wasted. Even the drier elements are utilized by insects, with hide beetles devouring any skin or tissue remaining on bones and micro-moths feasting on fur or feathers.

**Common burying beetle
(*Nicrophorus vespilloides*)**
Burying beetles are dedicated parents. They bury and transform small bird or mammal carcasses into underground nurseries, predigesting flesh and regurgitating it for the larvae.

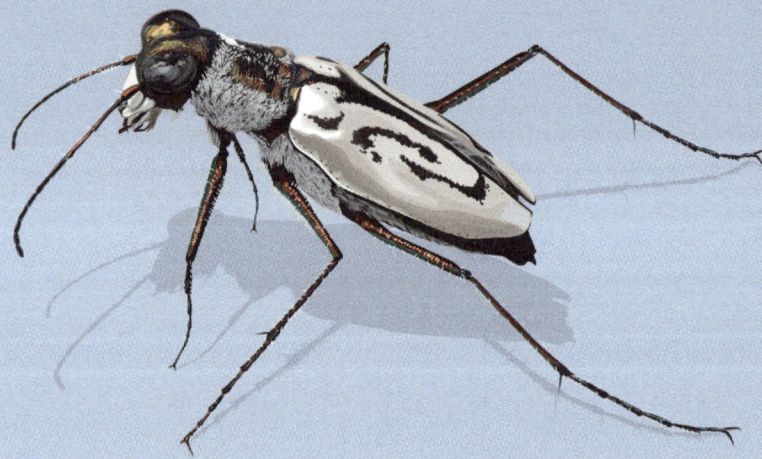

Running beetle
Certain ground-dwelling predators
have adapted perfectly to their terrestrial
lifestyle, such as the Australian tiger
beetle (*Cicindela hudsoni*), which holds
the record for the world's fastest running
insect to date. Its average speed is 5.6 mph
(9 kph), or around 171 body lengths
per second.

Jumping flea
The big leaps of the human flea (*Pulex
irritans*) are fueled by a special protein
called resilin. This is found in the fleas' legs,
and it functions as an elastic mechanism.
It is capable of storing, and subsequently
rapidly releasing, vast amounts of energy,
which propels the flea into the air.

Swarming larvae
Perreyia flavipes larvae move in a
rolling swarm. This behavior not
only provides protection, but also
allows them to move much faster
than they could as individuals.

The ways in which insects travel from place to place is incredibly varied, with numerous species exhibiting a wide range of movement beyond flight.

MOVEMENT

WALKING, RUNNING, AND JUMPING Many insects, such as ants and ground beetles, rely entirely on walking to cross various terrains. Flies, on the other hand, can climb and run across even the smoothest of surfaces, thanks to their "suction cup"-style feet. Fleas were once celebrated for their remarkable athleticism in flea circuses around the world. This was largely due to their incredible ability to jump up to 38 times their body size in only 1 millisecond. Grasshoppers and planthoppers are, likewise, great jumpers, relying on their powerful hind leg muscles to generate the force required for jumping.

SWIMMING AND SWARMING Dragonfly nymphs are entirely aquatic until they mature and transition to adulthood. During their time underwater, they use a unique form of locomotion: jet propulsion. By forcefully ejecting water through their anus, they create a jet of water that propels them forward, allowing them to move quickly to capture small aquatic prey. Pond skaters, sometimes called water striders, are able to effortlessly glide across bodies of water thanks to their specialized legs (covered in water-repellent hairs) and well-balanced weight distribution. These features allow them to make use of the water's surface tension to remain buoyant.

One of the most extraordinary ways in which insects move can be observed in the larvae of certain sawfly species, such as those of *Perreyia flavipes*, which is endemic to South America. These larvae are gregarious, meaning they gather in large groups, and move together in what is called a "rolling swarm."

Flight sequence
After taking off (A), an insect keeps flying through a sequence of four movements: backward, supination, upstroke, and pronation (B–E).

Key
- ● ● ● flapping direction
- ● ● ● wing rotation
- ← direction of wing movement

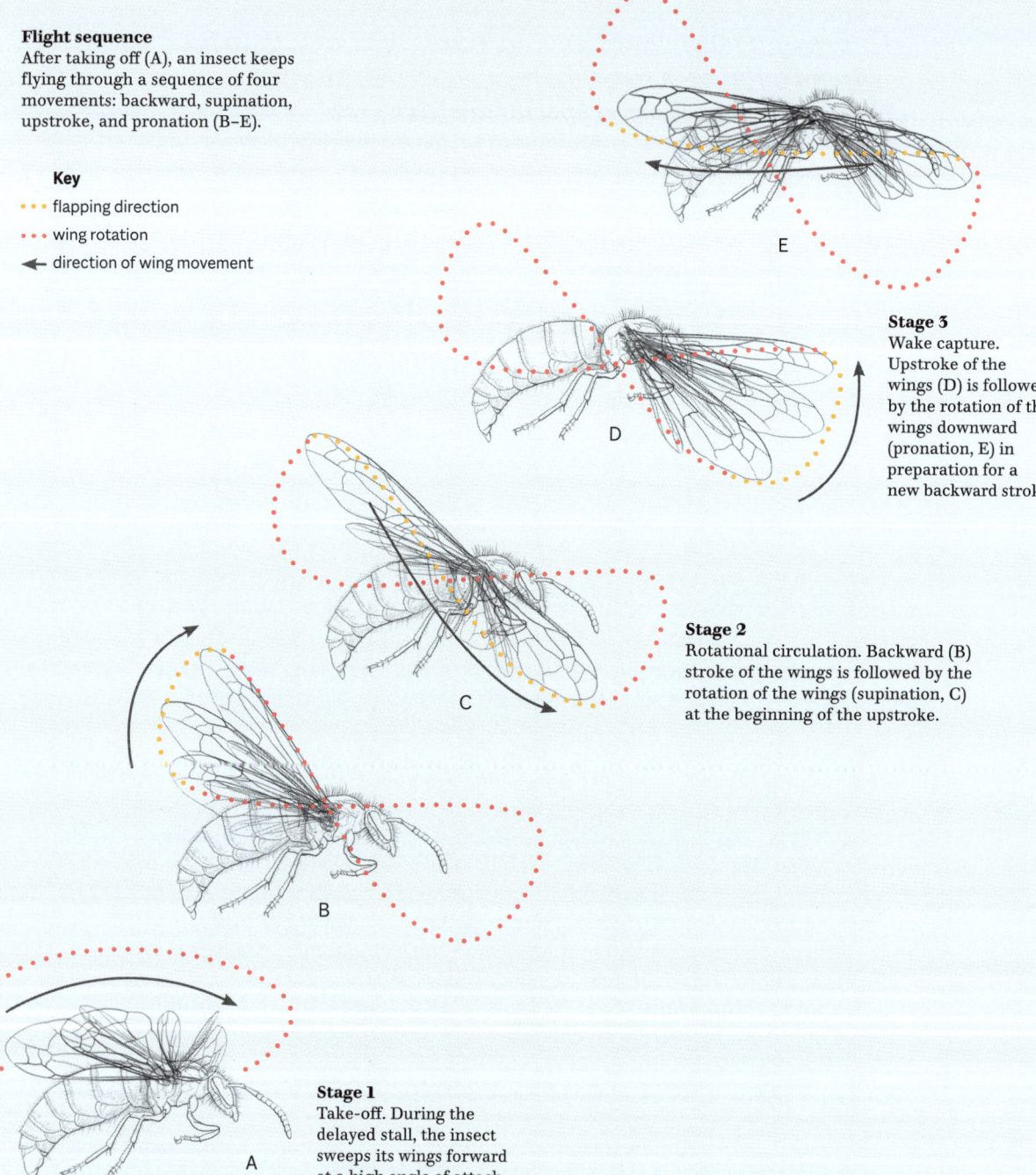

Stage 3
Wake capture. Upstroke of the wings (D) is followed by the rotation of the wings downward (pronation, E) in preparation for a new backward stroke.

Stage 2
Rotational circulation. Backward (B) stroke of the wings is followed by the rotation of the wings (supination, C) at the beginning of the upstroke.

Stage 1
Take-off. During the delayed stall, the insect sweeps its wings forward at a high angle of attack.

One reason for the success of insects is their ability to fly.
In fact, flying is so advantageous that only 5 percent
of all insect species are flightless.

FLIGHT

Insects were the first animals to take to the air. Their initial wings were more like extensions from the thorax or modified leg segments, and they helped maintain the insect's balance during impressive jumps and glides, or when skimming across the water to catch food or escape predators. The evolution of insects' muscles and wings was a highly significant development, enabling greater dispersion and long-distance migration. Remarkably, flying is also intrinsically linked to insect communication and reproductive strategies, such as the "waggle dance" performed by honeybees to communicate the location of food sources to other members of the hive, and the courtship displays of robber flies.

UNIQUE TECHNIQUE Insect wings are powered by two sets of robust muscles at the base of each wing, allowing for forward, backward, lateral, and hovering flights. The process of insect flight varies, but it generally comprises three key steps. The first is the delayed stall, and it is exclusive to insects. Fast, forward wingbeats at a high angle of attack (the angle between the oncoming air and the insect wing, 30 to 45 degrees) create an air vortex above the wings, draw the insect upward, and help with take-off. The second is rotational circulation, in which wings rapidly rotate backward, generating a backspin to initiate lift. The third is wake capture, where wings rotate before the return stroke, thus hitting previous turbulence to provide more lift and save energy.

AS FAST AS CHEETAHS Dragonflies are the fastest flying insects, and they can reach speeds of more than 31 mph (50 kph). Also remarkable are the reaction times of some insects. To evade predators, domestic flies, for example, can change direction in thirty-thousandths of a second. This is 1,000 times faster than humans perceive touch. In the air, large wings counteract air viscosity, allowing for low wingbeat frequencies and high speeds. For small insects, their size makes the air more viscous to navigate, thereby necessitating rapid wingbeat frequencies.

Insect migration is the most significant terrestrial movement of animals globally, and trillions of insects migrate thousands of miles every year.

MIGRATION

In the northern hemisphere, many insect species migrate toward higher latitudes in the spring, as flowers bloom progressively farther north. Their northward migration occurs over up to four or five generations, as each one takes advantage of the food resources available en route. In fall, as the temperature falls and days get shorter, a single generation migrates back south in one long journey. The insects never meet their parents or any experienced migrants, meaning their migratory behavior has an entirely genetic basis.

IMPORTANCE OF MIGRANTS Insects migrate primarily to increase reproductive output by breeding over as much of the year as possible. Additionally, they may migrate to escape predators or diseases. However, migrating insects play other important roles, such as redistributing pollen and essential nutrients. Many crop pests, such as locusts, are migratory species, but other insect migrants include species such as hoverflies that help control pest outbreaks. The world probably relies on insect migrants but just doesn't yet know it.

NAVIGATION Day-flying migrants have a time-compensated sun compass: they calculate their preferred route from the sun, even as it moves across the sky. Nocturnal migrants can migrate by starlight, and some are even able to detect Earth's magnetic field to use as a navigational tool. Insects migrate huge distances, and some can choose favorable winds to aid their journey.

WHICH SPECIES MIGRATE? Among insect migrants, the butterflies and dragonflies are well known. However, recent studies have shown that flies make up about 90 percent of migratory insects. Other insects such as ladybugs, lacewings, and bumblebees also migrate. The most abundant migrants, such as the cabbage white butterfly (*Pieris rapae*), can often be found in domestic gardens.

Marmalade hoverfly (*Episyrphus balteatus*)
This insect migrant is often found in gardens in Europe, North America, and North Asia. Its larvae play an important role as predators of aphid pests.

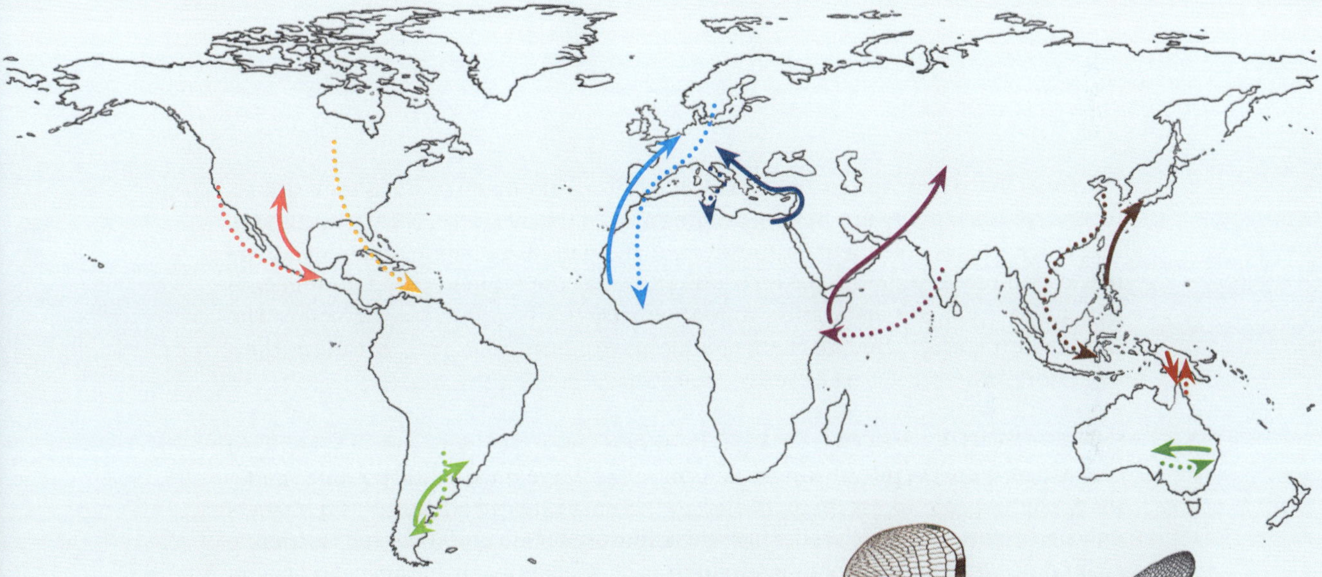

Key
Estimated routes of migration

— spring

••• fall

🔴 Monarch butterfly

🟠 American hoverfly

🟢 *Rhionaeschna bonariensis* (dragonfly)

🔵 Marmalade hoverfly

🔵 Bean seed fly

🟣 Globe skimmer dragonfly

🟤 *Calliphora nigribarbis* (blow fly)

🔴 Blue tiger butterfly

🟢 Bogong moth

Globe skimmer (*Pantala flavescens*)
This dragonfly follows the monsoon rains from India to Africa, staying in the rainy season so it can lay its eggs in predator-free puddles. It can make this ocean crossing in just a few days.

Most insects live solitary lives, but some, known as social insects, live together permanently in highly connected family-based colonies. These include ants and termites, as well as some bees, wasps, and aphids.

SOCIAL INSECTS

In insects, sociality is a spectrum, comprising solitary, communal, quasisocial, and eusocial (truly social) species. At one end of the spectrum, insects provide minimal care for their young; at the other, the eusocial species live in organized societies and have clearly defined roles. Organizing the workforce is central to social insect life. To keep the colony functioning, individuals have to do a complex array of jobs, such as rearing young, foraging, and defending themselves.

DIVISION OF LABOR For the sake of efficiency, social insects have evolved division of labor. Often, this is linked to worker age, with older workers doing more challenging jobs such as foraging. In some species, workers develop into different distinct physical types that have evolved to allow them to excel at certain jobs. To better defend the colony, soldier ants, for example, have a larger jaw, and soldier aphids have a larger body and tougher cuticle.

COMMUNICATION There are many differences between social insect species, but they all face the challenge of coordinating a large number of individuals and activities. To do this, social insects have evolved sophisticated communication, including the "waggle dance" language of honeybees. Honeybees returning to the hive pass on information about food sources through a figure-eight dance—the angle of the dance indicating direction, and its duration indicating distance. In contrast, scout ants lay complex pheromone trails that direct other worker ants to sources of food.

QUEENS AND KINGS A key aspect of the colony concerns which members reproduce. This varies, but typically colonies have only one individual, known as the queen, that lays eggs. In the ants, bees, and wasps, the queen is able to control

the sex of her offspring by choosing whether to lay fertilized eggs that develop into females or unfertilized eggs that develop into males. The queen's daughters are usually unable to mate. Instead of reproducing, they become workers, rearing the queen's offspring (their sisters and brothers) and keeping the colony running. Termite colonies (pp.52–53) have male and female workers, and the queen lives alongside a "king," who will mate with her throughout her life.

As we discover more about social insects, we recognize their profound ecological significance as they perform diverse roles, from pollination to pest control. Their impact is magnified by their sheer numerical dominance, particularly the ants (pp.48–49) and termites.

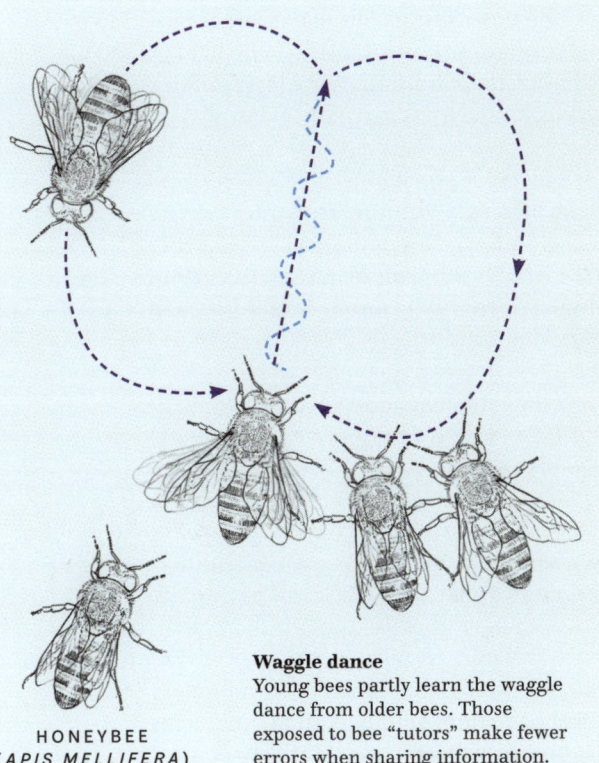

**HONEYBEE
(*APIS MELLIFERA*)**

Waggle dance
Young bees partly learn the waggle dance from older bees. Those exposed to bee "tutors" make fewer errors when sharing information.

Round dance
Recent research suggests that the round dance is a variation of the waggle dance, during which bees pass on information on the direction of food.

*Where there is one ant, there will always be more, because
ants live in colonies that range from a few dozen
individuals to many millions.*

SOCIAL INSECTS: ANTS

Social insects are those that live in colonies, working together to raise the young of just a few individuals. Among them, ants represent the largest and most diverse group. Recent estimates suggest there are around 20 quadrillion ants in existence—that is, 2.5 million ants for every person on Earth.

Currently, there are more than 14,000 known ant species ranging widely in size, living in arid deserts, grasslands, mangrove swamps, tree canopies, and underground. Some are specialist predators, fungus gardeners (pp.64–65), or scavengers, whereas others feed on honeydew, a sugar-rich liquid produced by aphids, scale insects, and other species with sucking mouthparts. These sweet-toothed species behave similarly to dairy farmers, protecting their aphid "cattle" and even moving them from plant to plant.

WHY ARE ANTS SO SUCCESSFUL? Ants are one of the most numerous and ecologically dominant insect groups on Earth. Leafcutter ants, for example, are serious pests of agricultural and forest crops in Central and South America. Some ant species have even become globally invasive. Red imported fire ants (*Solenopsis invicta*), originally from South America, have spread as far as New Zealand and, most recently, Southern Europe. The success of ants, in terms of their abundance across the world, is underpinned by their industrious and diverse exploitation of nesting and feeding opportunities and their flexibility in colony organization.

WHAT GOES ON IN AN ANT NEST?

A colony typically contains one or more queens (reproductively active females) and many closely related female workers. Workers do not usually lay eggs, and instead care for the queens' offspring. The males' only job is to find new queens, then mate. The workers forage, and also maintain hygiene in the nests. And when the colony grows, they dig out new chambers or scout for larger sites. Worker ants specialize in particular tasks, which may be linked to their body shape.

Supermajor and minor marauder ants (*Carebara affinis*) Despite the huge size difference, these ants are not only the same species, but also sisters from the same colony. The food provided to the worker larvae determines which form they develop into.

**Nest of the African paper wasp
(*Belonogaster juncea*)**
This African genus of social wasps builds
paper nests that are uncharacteristically
haphazard for a wasp. They rarely house
more than 10 individuals.

There is a lot more to the world of social wasps than being a nuisance. With more than 1,200 species, these insects exhibit huge diversity in the complexity, form, and functioning of their societies.

SOCIAL WASPS

WORKERS, QUEENS, AND MALES Picnic-bothering wasps are female worker yellowjackets (*Vespula*) looking for food to take back to the nest to feed their siblings. Their mother is the queen: the only egg layer in a society of 5,000 to 10,000 individuals. Early in the summer, the brood are all workers, but later in the season, the queen switches to producing males and gynes (next year's queens), which eventually disperse to mate. Males then die, and mated gynes look for somewhere to overwinter—garden sheds and attics are great choices because they are warm and well protected. In the spring, the new queens emerge and build their nests alone, using paper scraped from rotting wood or garden fences. Each queen provisions her brood with a diversity of insect prey. These offspring will be her first workers. Once they emerge as adults, the queen remains inside the nest as the dedicated egg layer.

A DIVERSITY OF NESTS Although all species of social wasps live in groups, there is huge diversity in the complexity of their colonies. The greatest diversity is found in the Neotropics (tropical region of Central and South America): more than 250 species belong to the Epiponini, or swarm-founding wasps, which establish new colonies as a swarm of hundreds of queens and workers. Their nests range from beautiful to bizarre. For example, *Apoica* is a genus of nocturnal paper wasps whose nests look like lampshades, and the nests of *Metapolybia* paper wasps resemble cowpats. Not all social wasps form large, complex societies. The *Polistes* species, which are found across the world, and the African *Belonogaster* wasps live in small groups of 10 to 50 individuals. The hover wasps of Southeast Asia live in groups as small as two or three individuals, in well-hidden nests that resemble lumps of earth. All group members in these simple wasp societies can reproduce, but most act as workers, helping raise the offspring of a single queen.

Termites are familiar as pests, causing significant damage to buildings and crops. However, they play an important role as "ecosystem engineers" and exhibit complex social organization skills.

SOCIAL INSECTS: TERMITES

With more than 3,000 described species, termites are descended from wood-feeding cockroaches, and some share similar features, including a reliance on symbiotic organisms in their gut to help break down cellulose in their woody diet. They live in permanent family groups (colonies), dividing tasks between the three castes: reproductives (fertile males and females), workers, and soldiers.

BUILDING A TERMITE COLONY Reproductives leave the nest as alates (winged form) in a synchronized swarm before pairing up, shedding their wings, and finding a site where they can begin their own colony. Worker termites are sterile and wingless, and they are usually pale and blind. Despite this, they are the backbone of the colony, responsible for foraging and nest building. Soldiers are larger than workers and are the defenders of the colony.

ECOSYSTEM ENGINEERS One of the most ecologically important developments in termite evolution is the ability to feed on soil. The workers of more than 1,000 species have highly complex and elongated guts to extract organic material from the mineral soil they ingest. Soil-feeding termites are so abundant in tropical habitats that they mirror the role of earthworms in temperate regions as "soil engineers," modifying the habitats where they live.

E. Workers of the Macrotermitinae subfamily forage above ground for plant material and are less soft-bodied than the subterranean workers of other termite groups. *Macrotermes* soldiers (below) are well armored, with enlarged heads and mandibles to protect the workers.

E

B. A complex ventilation system ensures that fresh air is circulated through internal air channels, maintaining a perfect temperature and humidity for the symbiotic fungus to grow.

A. Termite nests provide protection and a stable environment for their occupants. The largest and most complex nests are constructed by the Macrotermitinae subfamily, which have evolved to farm symbiotic fungi. The nest mounds of some *Macrotermes* species can be 26 ft (8 m) high, but many are simple and hidden underground.

C. Semidigested plant material is deposited onto a labyrinthine fungus garden to be fully broken down by the symbiotic *Termitomyces* fungus. The fungal mycelium and comb together provide a food source for the colony.

D. The royal chamber holds the queen termite and her king consort. The queen's abdomen is hugely swollen with eggs, and in some *Macrotermes* species, she can lay thousands of eggs each day.

Inside a beehive
Beehives contain an average of
approximately 40,000 bees during the
summer months, but this declines to
about 5,000 in the winter. Most
of the bees are workers.

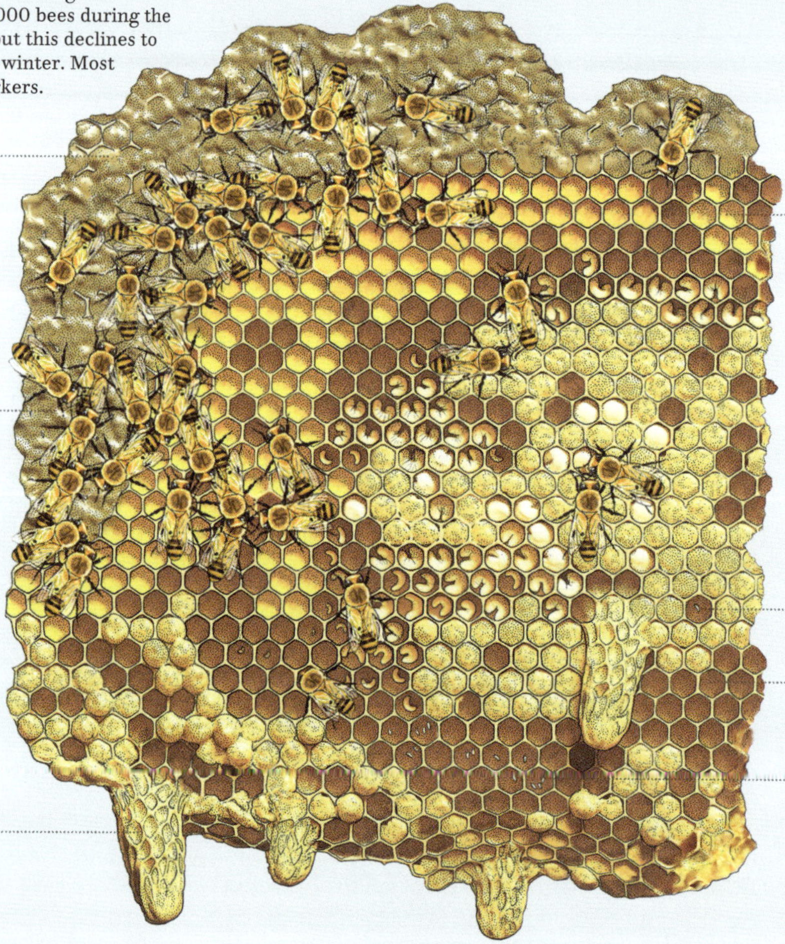

HONEY CELLS

POLLEN
CELLS

DRONE CELLS

SINGLE EGG
IN BROOD CELL

BROOD
CELL OPEN

MULTIPLE EGGS
IN SAME CELL
IMMATURE QUEEN

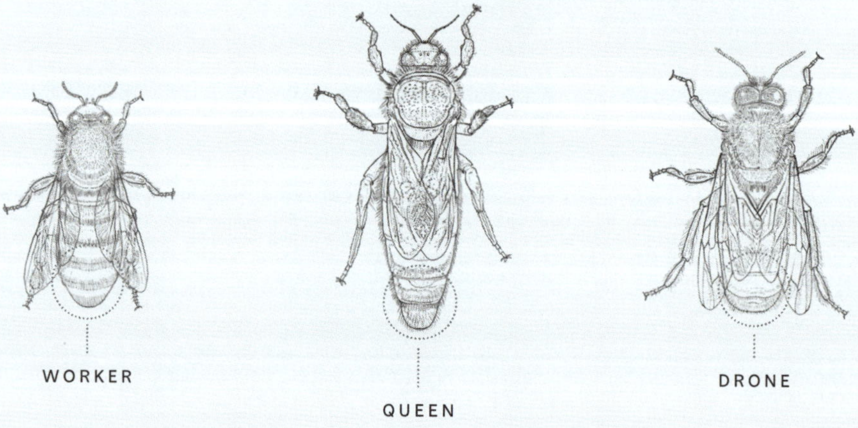

QUEEN CELLS

WORKER

QUEEN

DRONE

Social bee colonies work as a superorganism, with each bee performing a certain task—just like human cells working together to keep the body functioning.

EUSOCIAL BEES

Solitary bees make up about 90 percent of all bees. Single females lay eggs, provision them, and die before the larvae emerge. However, about 6 percent of bees are eusocial, building hives and living in organized societies. For bees and other insects to be considered eusocial, they must have reproductive division of labor, with only some individuals reproducing; cooperative brood care, with workers rearing the queen's young; and overlapping of at least two generations, meaning many of them technically still live with their mom.

BEE CASTES Caste can be determined by genes or by the quantity and type of food received as a larva. The main function of drones is reproduction, and they die right after mating. Workers are sterile and totipotent, meaning they can do anything from brood care and waste removal to nest rebuilding and foraging. The queen is the only one that mates and lays eggs. She outlives the workers—by how much depends on the species. In the eusocial sweat bee (*Lasioglossum umbripenne*), drones live 14 days, workers a month, and the queen up to a year. In western honeybees (*Apis mellifera*), drones live eight weeks, workers one and a half to five months, and the queen between two and six years.

ADVANTAGES OF EUSOCIALITY Each bee has a specialized role within the colony. Some workers focus solely on brood rearing, thus improving offspring care. Others are responsible for foraging. Eusocial bees are also capable of coordinating nest defenses chemically or by stinging and biting predators or parasites. And corbiculate bees (bumblebees, honeybees, and stingless bees) are able to deal with poor weather conditions and food shortages by regulating the temperature of their colony and storing food in cells (honey). In winter, when temperatures are too low to forage, they use stored resources to nourish the queen and her brood, thus keeping the colony functioning.

There are few things more impressive in nature than the structures produced by social insects, such as ants, termites, and wasps. Their creativity, complexity, and variety are simply spectacular.

SOCIAL INSECT-BUILT STRUCTURES

Social insects build nests for protection from predators, disease, and the climate; to produce food; and to facilitate reproduction. There is, therefore, huge variability in the types of nests they construct. Wasps build nests to last one summer, whereas termites can build them to last decades—and even 2,000 years in one example from central Africa.

FARMING FUNGUS Fungus-growing termites are farmers. They cultivate fungus within complex mounds that can be several yards tall (pp.52–53). These are predominantly conical, made from hardened soil produced by salivary excretions. They comprise a complex network of pathways, plus ventilation tunnels and pores to regulate temperature and humidity. This creates the ideal environment for fungus gardens, where the termites farm the specific fungus they eat, using semidigested plant material collected from outside the mounds.

These termite mounds make a positive impact on their surroundings. Their nutritious soil enables them to sustain diverse and unique plant communities, and they provide nest sites for birds, bats, reptiles, and invertebrates, which are likely using the termite nest to gain the same benefits as the termites (a protected and climatically regulated nest space) without any of the work. In fact, there is a specific word for invertebrates that cannot survive outside of termite mounds—termitophily—and this association has evolved independently at least 40 times, with the earliest examples dating back 100 million years.

BUILDING BIVOUACS
Army ants are unique among ants because they do not make permanent nests using natural materials. Instead, they build nests, known as bivouacs, solely from their own interconnected bodies. These bivouacs can contain thousands of individuals, with each ant attaching to another (using their claws and mandibles) to form the nest walls. The queen is protected in the center of the nest, which is often less dense than the thick exterior of writhing bodies. Nomadic army ants may build a new bivouac each night, sometimes even continuing to travel by coordinating their movements.

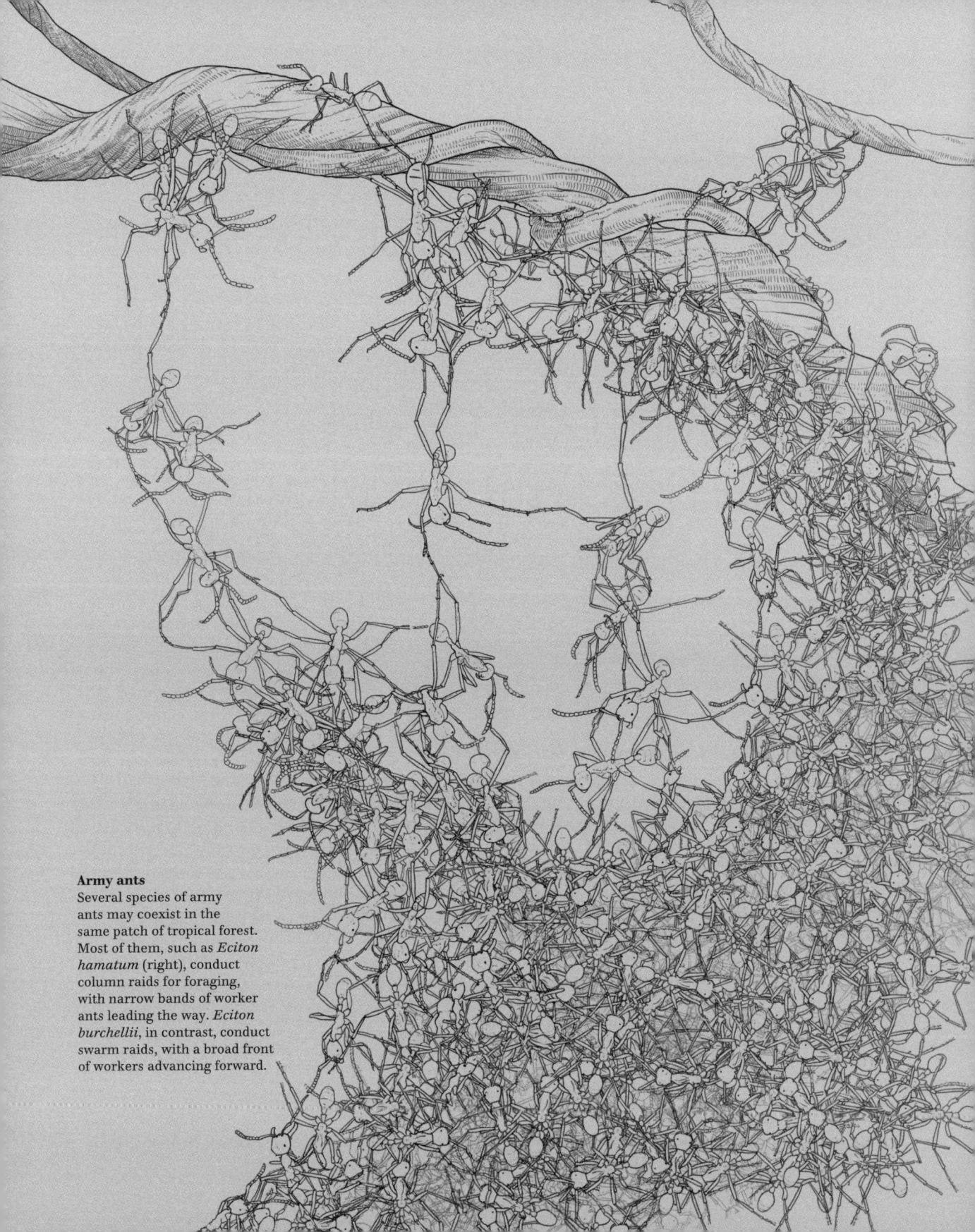

Army ants
Several species of army ants may coexist in the same patch of tropical forest. Most of them, such as *Eciton hamatum* (right), conduct column raids for foraging, with narrow bands of worker ants leading the way. *Eciton burchellii*, in contrast, conduct swarm raids, with a broad front of workers advancing forward.

Insects are naturally equipped to build structures using silk made in their bodies, and underwater caddisfly larvae are particularly adept at this.

SILK-BUILT STRUCTURES

Insect builders need three basic skills: finding, measuring, and joining building blocks. They use silk to join materials together, which is made in the body as a liquid, then turned into a thread when extruded. Although often associated with spider webs, silk is made and used in this way by many insect species, including caddisflies. A caddisfly larva touches an object with its silk extruder, then draws a thread and touches another object to join the two together. It repeats the process until it has built a shelter. Larvae in many insect groups, but especially caddisflies and moths, disguise or protect themselves by creating portable cases into which they retreat if threatened. Although small, these structures can be so numerous that they affect water flow and the microhabitats of streams and rivers.

BUILDING MATERIALS The natural world is replete with potential building material for insects, such as sand grains, live and dead leaves, and pieces of wood. Many objects will be too big or too small, but special flexor nerves in the caddisfly larva's joints and body segments enable it to make measurements. It can measure out the right-sized piece of plant or wood from a larger section, then cut out the bit it wants with its powerful mandibles (the largest components of its mouthparts).

CATCHING PREY Insects use silk to build structures for reasons other than protection. The larvae of some caddisfly species make a fixed tubular shelter on a rock or plant in which they hide. This is attached to carefully constructed food-catching apparatus. Other species wait for prey to become entangled in their web of silks. Species in the Philopotamidae family of caddisflies make and live in a silk net so fine that it can trap tiny phytoplankton (microscopic plantlike organisms).

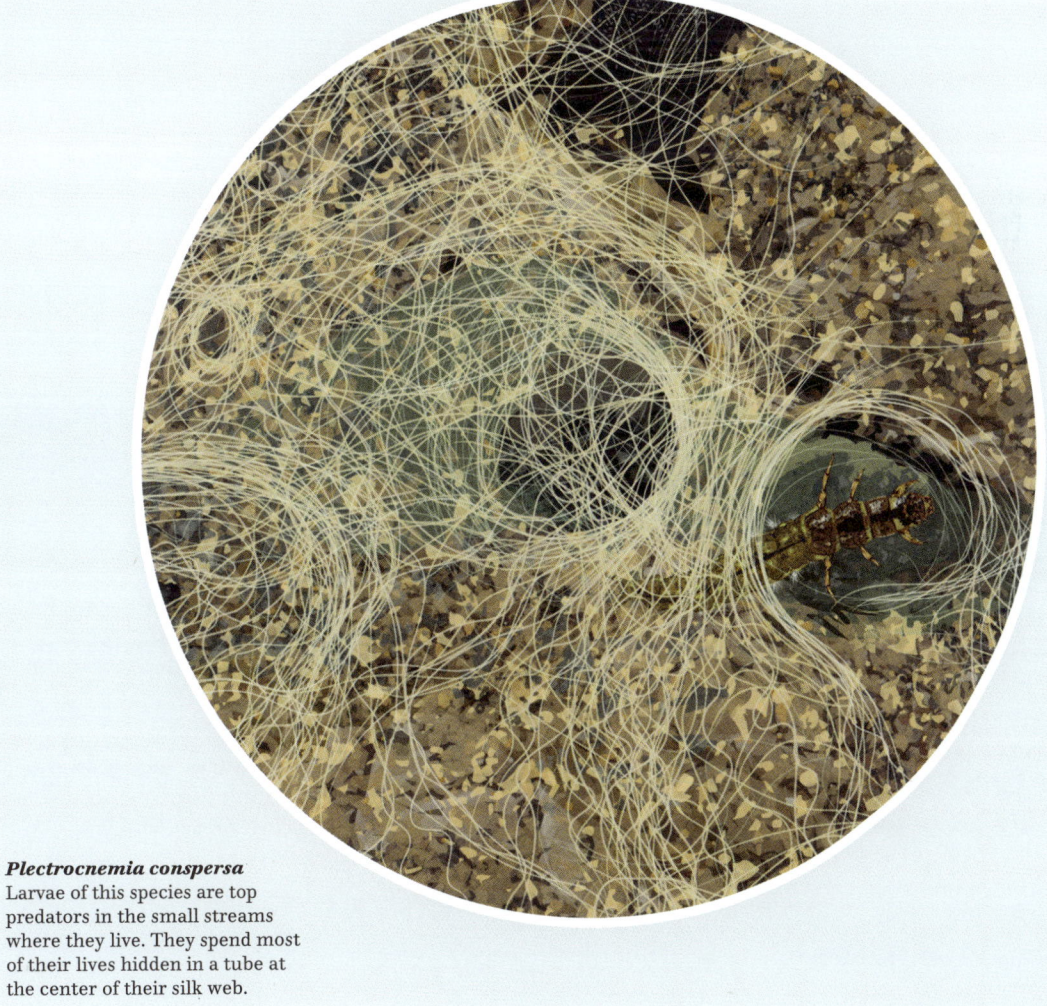

Plectrocnemia conspersa
Larvae of this species are top predators in the small streams where they live. They spend most of their lives hidden in a tube at the center of their silk web.

Orange-tip butterfly (*Anthocharis cardamines*)

INTERACT

CHAPTER III

HOW DO INSECTS alter the world around them? Some cultivate harmonious mutually beneficial relationships with plants and animals, while others show hostile behaviors as parasites, predators, or pests. Uncover a diversity of pollinators, medicinal maggots, specialists, and generalists that make a huge impact on the natural world.

The majority of insects encounter other insect species at some point during their life cycle. The nature of these encounters varies and may be positive, negative, or neutral for either or both insects.

INSECT–INSECT INTERACTIONS

MUTUAL BENEFITS Mutualistic insect–insect interactions are those in which both parties benefit. An example of a (mostly) mutualistic interaction is the "farming" and guarding of plant-feeding mealybugs, such as *Formicococcus njalensis*, by ants such as *Crematogaster* species. The ants protect the mealybugs from attacks by predators and parasitoids while consuming the sugary secretion (honeydew) excreted by the mealybug.

PREDATION AND PARASITISM Insect–insect interactions sometimes benefit one insect at the expense of the other. An example of this is predation, such as when a long-legged fly captures, kills, and eats a soft-bodied insect like an aphid. Another is parasitism, when one insect appropriates the body or resources of another (pp.68–69). This could be a parasitoid wasp laying eggs inside a host such as an aphid, or a dotted bee-fly (*Bombylius discolor*) laying eggs in a solitary bee's nest, where the bee-fly larvae will consume the resources stored there by the bee for her own offspring.

CASCADING INTERACTIONS Insect–insect interactions can have implications not only for the insects, but also for other plants and animals in the ecosystem (pp.186–187). Competition between two pollinator species (such as a bumblebee and a hoverfly) on a flower can influence how they forage for food and move between plants. This may affect the plant's seed set (determining the seed number and mass) and its success in the habitat.

AGGRESSIVE BEHAVIOR Sensory cues can be important in insect–insect interactions. In particular, visual and chemical cues displayed by one insect can influence the behavior of another, such as by increasing or decreasing aggression between species. Damselflies in the genus *Hetaerina*, for example, engage in aggressive territorial behavior toward members of other species, including chasing, ramming, or grabbing intruders. In fact, males of all species in the genus defend mating territories (small areas of flowing water), which females visit to mate and lay their eggs on submerged plants. Male rubyspot damselflies have a red spot on their wings that contains sufficient information to distinguish their sex and even age, ultimately triggering aggression between sexually mature male damselflies.

Ants tend mealybugs on a cocoa tree in West Africa
Mealybugs are small, sap-sucking scale insects. They are covered in white waxy powder, and some carry a plant virus that they transfer to cocoa trees when feeding. This virus causes swollen-shoot virus disease, which has a devastating effect on cocoa crops.

Fascinating interactions between ants and fungi play important roles in shaping ecosystems, contributing to the recycling of nutrients, and have even influenced the evolution of ants and fungi.

ANTS AND THEIR FUNGUS GARDENS

NURTURE Leafcutter ants lack a specialized digestive system for breaking down cellulose, a major constituent of plants, so they cannot consume plant material directly. Despite this, around 50 species of leafcutter ants dominate plant consumption in Central and South America, using a reciprocally beneficial interaction known as a mutualism. Leafcutter ants have formed a mutualism with a spongy fungus, *Leucoagaricus gongylophorus*, referred to as the "fungus garden." The ants supply plant fragments to the fungus, and in return consume fungal tissues called staphylae. Staphylae contain nutrients, and also specialized enzymes that pass through the ant's digestive tract onto plant material, transforming previously indigestible plants into accessible nutrients. Some nutrients are used by the fungus to continue growth, while others create nutrient-rich food rewards, through structures called gongylidia, for the ants to consume.

MANIPULATION This mutualism is threatened by specialized parasitic fungi (*Escovopsis*), which infiltrate the fungus gardens and siphon off nutrients for their own gain. As the parasitic fungi mimic the scent and texture of food rewards produced by the fungus garden, ants continue to "feed" them until their delicate agricultural system begins to collapse. Once the alarm is raised, the ants rapidly identify and remove infected patches in their garden to prevent further damage. A more sinister parasite is *Cordyceps* fungus, which hijacks the bodies of its ant victims to spread spores. These attach to the ant exoskeleton, then penetrate its body, after which the fungus consumes the ants' tissues. Vital organs are left intact to keep the ant alive for as long as possible until eventually the ants' nervous system is taken over. Infected ants exhibit abnormal behaviors: climbing to higher positions on vegetation, then dying. This bizarre last act enables the fungus to spread its spores by wind to continue its cycle of infection.

ANT/FUNGUS MUTUALISM

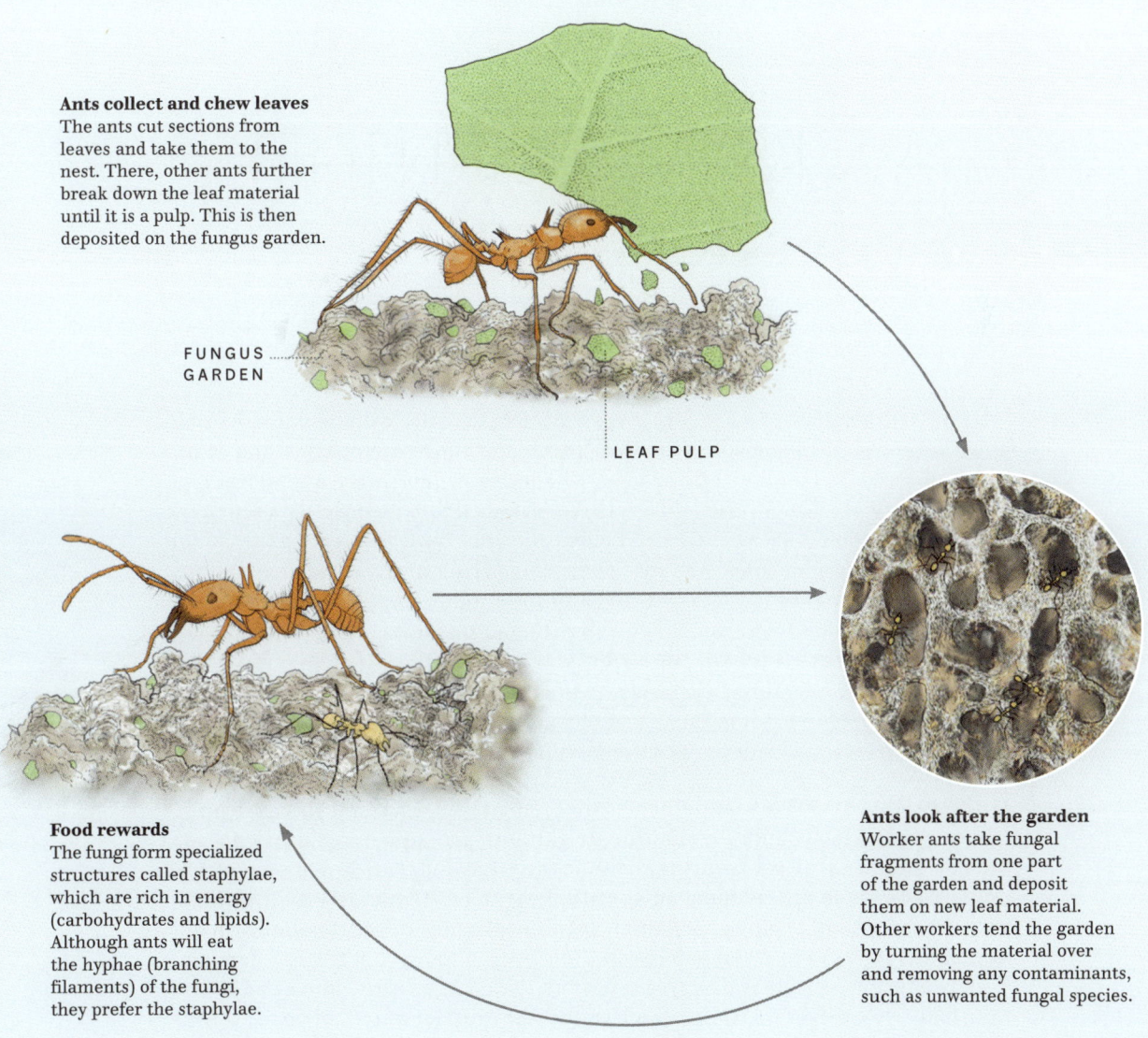

Ants collect and chew leaves
The ants cut sections from leaves and take them to the nest. There, other ants further break down the leaf material until it is a pulp. This is then deposited on the fungus garden.

FUNGUS
GARDEN

LEAF PULP

Food rewards
The fungi form specialized structures called staphylae, which are rich in energy (carbohydrates and lipids). Although ants will eat the hyphae (branching filaments) of the fungi, they prefer the staphylae.

Ants look after the garden
Worker ants take fungal fragments from one part of the garden and deposit them on new leaf material. Other workers tend the garden by turning the material over and removing any contaminants, such as unwanted fungal species.

Some bee species have ditched the hard work of motherhood and found a way to get others to build their nests and feed their offspring.

BROOD PARASITISM IN BEES

Most bees are solitary nest-building insects that collect nectar and pollen for their offspring. The egg then feeds on its own pollen ball. Brood parasitic bees, in contrast, take advantage of other bee species to rear their offspring. These insects do not have pollen-collecting structures. Instead, they have elongated mandibles and knightlike armor, and they are perfumed to smell like their hosts. Some adult females produce chemicals similar to those produced by the host bees to mask their smell and enter host nests undetected. In some nomad bee species, males apply secretions to females during mating, which makes them smell like the chemicals the host uses for nest marking. Other brood parasitic bees chew leaves and spread them onto their body to mask their odor.

CLEPTOPARASITIC BEES LAYING IN CLOSED CELLS These bee species find a sealed cell in which to lay an egg. The larvae are hospicidal, meaning they are in charge of killing the host's offspring. They develop quicker than the host larvae and are highly aggressive. Sometimes, several adult females parasitize the same cell, so hospicidal larvae have to kill other brood parasites. The interaction between *Dioxys pomonae* and its host *Osmia nigrobarbata* is an example of this type of brood parasitism.

CLEPTOPARASITIC BEES LAYING IN OPEN CELLS This type of parasitism is the most common among bees. The bees find a nest that is still being provisioned and lay their eggs inside the wall of the host cell or inside the pollen ball, unnoticed by the nest-building mother. Dark bees (*Stelis*) interact with their hosts, mason bees (*Osmia*), in this way. The fifth-instar larva of the dark bee punctures the mason bee larva with a bite and drinks its contents.

ADULT, CLOSED CELL PARASITISM

BLACK ORCHID BEE CELL
(*EULAEMA NIGRITA*)

1.

The black orchid bee is the host bee. The female builds a cell in a nest constructed from excrement, mud, and resin. Black orchid bees usually nest in tree cavities or underground.

2.

The female lays her eggs in cells that she has provisioned, then seals them closed to protect her brood. Once the host female has provided enough sustenance for all her eggs, she dies.

3.

The parasitic emerald cuckoo orchid bee looks for a host nest, inspecting cracks and holes in the soil and also in bricks and trees.

4.

Inside the host nest, the parasitic bee searches for a suitable cell to parasitize.

5.

Once the parasitic bee finds a cell that has been provisioned by its host, she kills the host egg or larva. She makes a small hole in the cell wall above the level of the provisions, deposits her egg through the hole, then reseals the opening.

EMERALD CUCKOO ORCHID BEE
(*EXAERETE SMARAGDINA*)

6.

The emerald cuckoo orchid bee leaves her offspring in the host bee's nest, having avoided the laborious processes of nest building and provisioning.

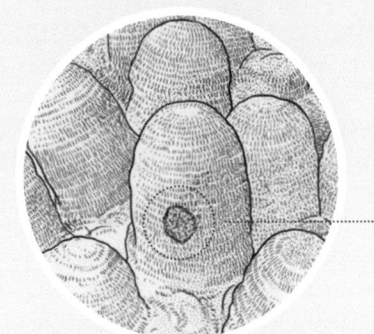

PARASITIZED NEST

Parasitoids are insects whose offspring develop by feeding in or on the bodies of other insects, eventually killing them.

PARASITOIDS

Parasitoids begin life as parasites living within a host, but, ultimately, they consume their host, so they resemble predators killing their prey. Parasitoids have evolved several times, but most are very small to medium-sized wasps, or they belong to a family of flies (Tachinidae) that look a little like houseflies. All parasitoid wasps descended from a single ancestor that lived about 250 million years ago. Today, there are at least 1 million parasitoid species, probably many more, and all are abundant within terrestrial ecosystems.

CATERPILLAR VICTIMS The females of a parasitoid wasp called *Cotesia glomerata* target and attack the caterpillars of the large white butterfly (*Pieris brassicae*). Only a few millimeters in length, the wasp detects the caterpillar using chemical gases known as volatiles. These are produced by the host plant and from the caterpillar's feeding activities and excreta. Many parasitoids are solitary, with only one egg developing in each host, but *C. glomerata* lives in groups, and the female injects 30 to 40 eggs into the butterfly caterpillar. Along with her eggs, she injects a virus, which hijacks the host's cellular machinery to make the caterpillar produce proteins that disable its immune system and therefore prevent it from attacking the developing parasitoids. Initially, the parasitoids feed carefully, keeping themselves alive and allowing the host to grow. When the caterpillar is nearly full size, the parasitoids consume all the host tissues from the inside and emerge through the caterpillar's skin to pupate in a ring around its empty husk. But these parasitoids don't always get their own way. There are further species of wasp that specifically target *C. glomerata* and lay their eggs inside them. These species, parasitoids of parasitoids, are called hyperparasitoids.

Parasitoid wasp larvae and caterpillar
A small parasitoid wasp (*Cotesia glomerata*) laid
her eggs within this caterpillar (*Pieris brassicae*).
The caterpillar was killed by the developing brood
of parasitoid wasp larvae, which burst through its
skin prior to pupation.

PARASITOID WASP
(*COTESIA GLOMERATA*)

Female egg parasitoids lay their eggs inside the eggs of other larger insects. For parasitoid wasps, a single host egg is sufficient to provide all the resources the wasp needs to develop to pupation.

EGG PARASITOIDS

FINDING A HOST The task of finding a small egg in a complex environment is daunting for a tiny egg parasitoid. Some species do so by locating an adult female, then following her until she oviposits (lays eggs). Often, they ride shotgun on the female, only jumping off when she lays. After the egg parasitoid has oviposited into the other insect's egg, the existing embryo dies and is consumed by the developing egg parasitoid larvae.

SPECIAL ADAPTATIONS Fairyflies of the family Mymaridae are tiny egg parasitoids that attack the eggs of many types of insects. A few species target the eggs of aquatic insects, such as water beetles. To help them in this environment, the fairyflies' wings are greatly reduced and oar-shaped. The wing vein provides the shaft, and the blade is formed from a small area of wing membrane with long hairs. These egg parasitoids "paddle" through the water, and they can spend long periods submerged. They are so small that they can obtain oxygen simply through diffusion across their body surface.

BIOLOGICAL CONTROL By killing the eggs of pest insects, egg parasitoids can be very effective biological control agents. For example, wasps from the family Trichogrammatidae are available commercially for release in orchards to control the codling moth (*Cydia pomonella*), which attacks the fruit of apple trees. Codling moth caterpillars cause damage by tunneling into apples, but the wasps oviposit into codling moth eggs and kill the developing embryo before it is able to hatch. There is then no need to spray insecticides on apples intended for human consumption. Biological control companies have developed facilities to mass rear very large numbers of wasps, which are sold to orchard owners around the world at times when the pest moth is active.

LIFE CYCLE OF *TRICHOGRAMMA* WASP SPECIES

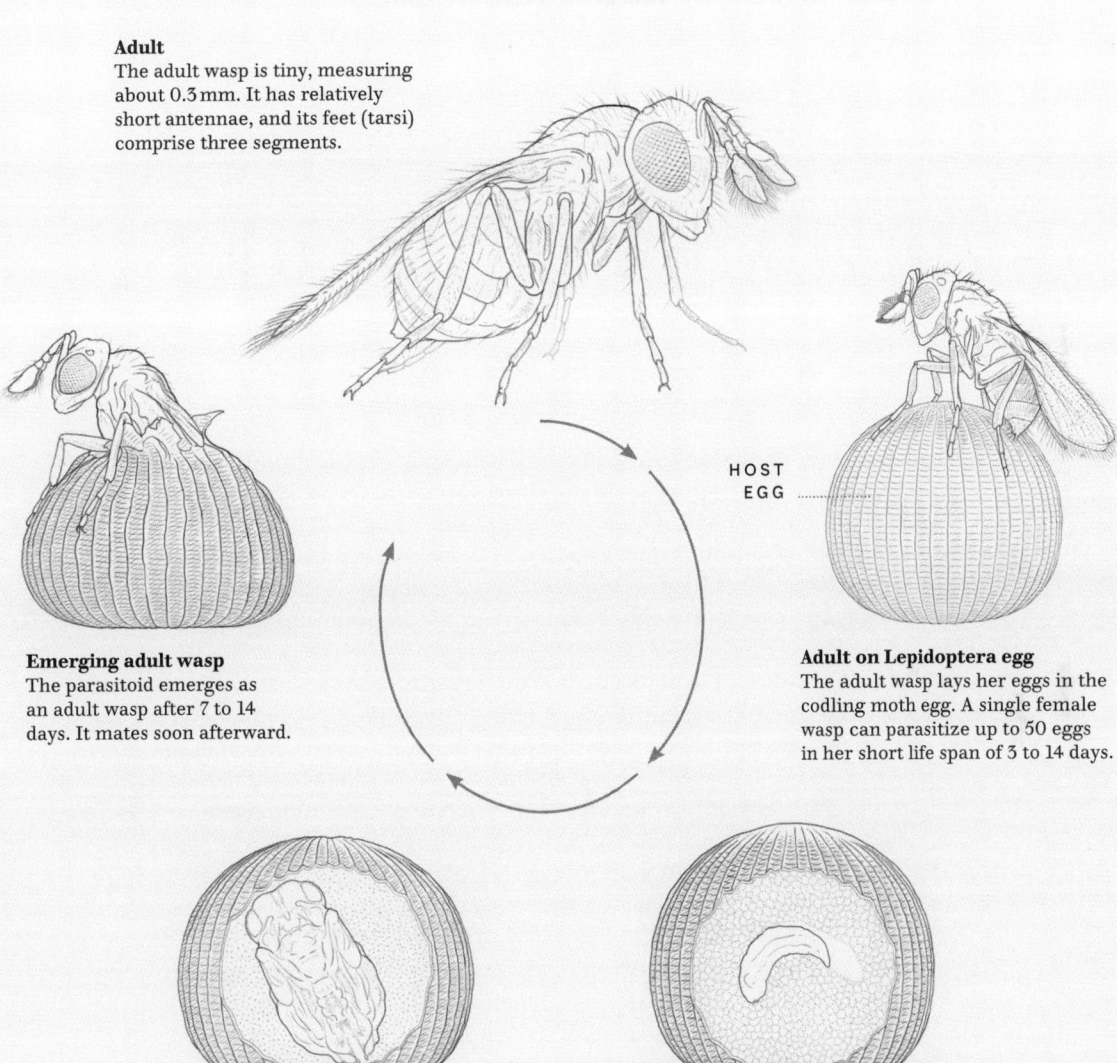

Adult
The adult wasp is tiny, measuring about 0.3 mm. It has relatively short antennae, and its feet (tarsi) comprise three segments.

HOST EGG

Emerging adult wasp
The parasitoid emerges as an adult wasp after 7 to 14 days. It mates soon afterward.

Adult on Lepidoptera egg
The adult wasp lays her eggs in the codling moth egg. A single female wasp can parasitize up to 50 eggs in her short life span of 3 to 14 days.

Pupa in Lepidoptera egg
The parasitoid wasp larva pupates within the host egg.

Larva in Lepidoptera egg
The wasp larva hatches within the host egg and consumes its contents.

In the beetle world, the relationship between the hunter and the hunted (predation) is just as exciting as it is in the mammalian world, featuring ambush predators, big game stalkers, and generalist scavengers.

PREDATORY BEETLES

Carabids are a type of predatory beetle, of which there are more than 40,000 species throughout the world. They have a variety of grisly hunting behaviors but are mainly generalists, eating any prey they come across. This means they are useful on farmland, controlling a range of crop pests from aphids to weevils.

KEEN-EYED HUNTERS Big-eyed beetles (*Notiophilus biguttatus*) live up to their name. Their huge, bulbous eyes have many more lenses, known as ommatidia (pp.16–17), than other carabids. They use their fantastic vision to stalk springtails and other invertebrates, sneaking close enough to attack before the prey can escape. But springtails have developed an amazing way to leap to safety. They have a specialized jumping limb, like a tail, which curls beneath them. When threatened, they unfurl it with force and launch themselves away.

CAPTURE SPECIALISTS Hair-trap ground beetles (*Loricera pilicornis*) use a different strategy. Their antennae have special long hairs that sweep over the ground, trapping springtails like a net before they can jump away. These hairs (setae) are around 10 times stronger than other hairs on the beetle's body. Similarly, Prussian plate-jaw beetles (*Leistus spinibarbis*) have large, flat jaws with bristly protrusions that trap their bouncy prey.

BIG GAME HUNTERS Tiny, fast prey is one thing, but bigger game presents different problems. Carabids such as the *Carabus* and *Tefflus* species track slugs and snails by sniffing out and following their mucus trails. Snail hunters are then presented with the challenge of armor. Some carabid species use their large jaws to break open snail shells, while others have slender heads and jaws and crawl into the shells themselves to feed.

**Big-eyed beetle
(*Notiophilus biguttatus*)**
These beetles use their 200-degree field
of vision to stalk prey before striking
with deadly accuracy at close range.

Lacewing feeding on an aphid
Green lacewing larvae are
predators of aphids and other
similar insects, many of which
are pests of agricultural or
horticultural crops.

Farmers and gardeners around the world benefit from the many different types of insects that feed on aphids. These natural enemies can be used as biological control agents.

APHID HUNTERS

Ladybugs are one of the most well-known groups of aphid hunters, with adults consuming thousands of aphids in their lifetime. Aphids are also prey for many other predatory beetles, hoverflies, and lacewings. Green lacewings (*Chrysoperla carnea*) are found worldwide in many different habitats, including gardens and agricultural fields. Adults feed on the sugar-rich nectar and pollen resources provided by many plant species, but their larvae are predators that actively hunt for aphids and other soft-bodied invertebrates, such as thrips. They locate their prey predominantly using complex chemical cues, whereby they identify potential prey patches from the chemicals emitted by both the aphids and the plants they are feeding on. Once lacewing larvae have located an aphid, they lunge forward to impale or grip it with their hollow, sickle-shaped mouthparts. Enzymes are then injected into the prey through these mouthparts, which digest the internal body contents of the aphid and enable the lacewing larvae to suck them into their mouthparts for consumption. Each lacewing larva can consume more than 100 small insects per day throughout their lifetime.

BIOLOGICAL CONTROL AGENTS Understanding the interactions between aphids and aphid hunters is essential if these natural enemies are to be used within biological control programs. It can help determine the most effective strategies for deploying predatory insects, including how many predators are needed to achieve optimal pest control. Understanding the behavior of aphid hunters and how they alter their foraging strategies as the density of their prey changes is important to maximize pest control. Lacewing larvae, for example, increase their rate of predation as the density of aphids increases, until they are satiated and begin to capture fewer aphids. The effectiveness of aphid hunters as biological control agents is influenced by other factors, too. For example, ladybugs leave behind a trail of chemicals, which can deter other insect predators.

Plants and the insects that eat them have evolved alongside one another for millions of years and together account for more than half of all species on Earth.

INSECTS AS HERBIVORES

FEEDING STRATEGIES Most species of insect are herbivores, and they have evolved to consume every conceivable plant part. Possibly the most familiar insect herbivores are those that chew plant foliage, leaving holes or scars that remain long after the insect has left. Caterpillars of some moths, such as the witchetty grub (*Endoxyla leucomochla*), feed while hidden inside stems, branches, and roots, where they are protected from most predators. The pea aphid (*Acyrthosiphon pisum*) and other true bugs (pp.160–161) remain on the plant's surface, sucking fluids through their piercing mouthparts.

PLANTS FIGHT BACK The damage caused by insect herbivores will rarely kill a plant outright, but it can often harm them. As a result, plants have evolved to protect themselves from herbivores either physically—using hooks or hairs, for example—or by deploying a wide variety of toxic chemicals within their tissues. Particular insect species have evolved to overcome the defenses of individual plant species. Most insect herbivores are, therefore, rather specialized, feeding on one or a few closely related plant species or on unrelated plants that are chemically similar. Important and vulnerable plant parts, such as seeds, tend to be particularly well defended chemically, but insects such as the bean weevil (*Acanthoscelides obtectus*) have evolved to cope with the chemicals. Many insects have also turned plant toxins to their advantage, retaining them within their bodies and advertising their own unpalatability to predators via vivid coloration.

GIVING PLANTS A CHANCE

Insect herbivores may harm the growth and reproduction of individual plants, but they also have a role in maintaining high plant diversity in habitats such as tropical rainforests. Where a particular plant species reaches high density, it is more easily detected and exploited by insect herbivores. This can reduce the abundance of the plant, thereby giving rarer plant species—which are harder for the insect to find— a competitive advantage and allowing a greater diversity of plant species to coexist.

Orange-tip butterfly (*Anthocharis cardamines*)
In some species of moth and butterfly, including the orange-tip butterfly, caterpillars that initially harm a plant (in this case, lady's smock or cuckoo flower, *Cardamine pratensis*) by consuming the leaves and seeds later become beneficial pollinators of the same plant as adult insects.

EGG

Orange-tip caterpillar
Although mainly herbivorous, these caterpillars will cannibalize eggs and other larvae if they meet them on the same plant.

More than 1 million species of insects have been identified globally, but only a very small number are considered to be "pests" or threats to global crop production.

PESTS ON CROP PLANTS

DIRECT AND INDIRECT DAMAGE Pest species damage crops that farmers grow by reducing yields and/or lowering the quality of the harvest. Direct damage often occurs when herbivorous insects feed on crops, but some damage is indirect, such as when insects act as vectors and transfer disease between plants. The tobacco whitefly (*Bemisia tabaci*), for example, causes some direct damage, but also transmits viral diseases to more than 800 plant species, including tomato, cotton, and sweet potato.

BAD FOR BUSINESS Often, it is the immature insect that damages the crop, but sometimes it is the adult, and occasionally it is both. For example, the cabbage stem flea beetle (*Psylliodes chrysocephala*) adult feeds on the leaves of oilseed rape plants, while the larvae of this species tunnels into the leaf stalks and plant stems. Some insect pests cause problems below ground. The large bulb fly (*Merodon equestris*), for example, damages the bulbs of several plants, notably daffodil species. It has spread across the world with the bulb trade since the 18th century and is estimated to spoil up to 20 percent of commercial crops. Some insect pests can cause very serious problems for farmers and growers, including the grape phylloxera (*Daktulosphaira vitifoliae*). This species was accidentally introduced from North America to Europe in the 19th century and went on to destroy more than two-thirds of European vineyards.

Many crop pests are not effectively controlled by natural enemies, such as predatory and parasitic insects, particularly when the environment is not managed to maintain a good diversity of species or when chemical pesticides are overused. Farmers often seek to improve their yields by growing pest-resistant or higher-yielding crops varieties (pp.220–221).

STAGES OF CODLING MOTH

Larva tunnels into the fruit, leaving a red-ringed entry hole
Initially, feeding results in a small cavity just beneath the surface of the skin of the fruit.

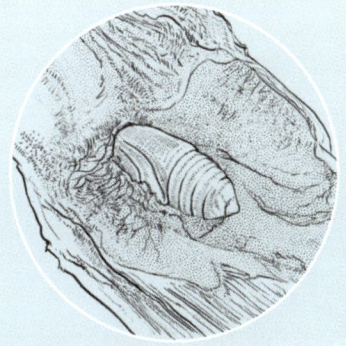

Larva pupates
The larva spins a cocoon in a crack in the tree trunk and pupates inside. In Northern Europe, there is a single generation, but in warmer parts of Europe, there are two or more generations.

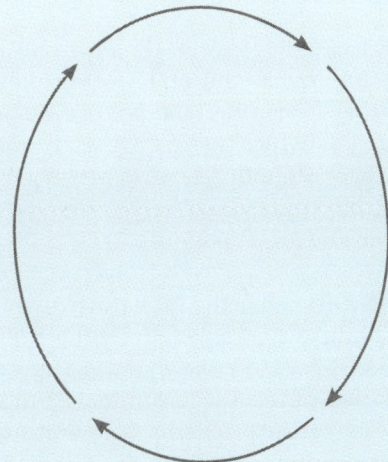

Larva tunnels down to the core
At the core of the fruit, the larva may feed on the seeds and fill the cavity it has created with frass (excrement).

Pinkish-white larva grows to 3⁄4 in (20 mm) in length
Larval development takes about four weeks to complete, at which point the larva leaves the fruit.

Pine beauty moth (*Panolis flammea*)
By the 1970s, the aptly named pine
beauty moth was a serious pest of
lodgepole pine in Scotland. The adults
fly in March and April, and the larvae
emerge in May to feed on young pine
shoots and needles.

Naturally growing forests—resilient ecosystems with a diversity of trees and other plants—rarely experience pest outbreaks, but managed forests, with single species of trees, sometimes suffer insect damage.

FOREST PESTS

FOLIAGE-FEEDING INSECTS One of many insects that feed on the foliage of trees, the pine beauty moth (*Panolis flammea*) occurs naturally across much of Europe. In Scotland, for example, it feeds naturally on the needles of Scots pine (*Pinus sylvestris*), but only in low numbers. However, when lodgepole pine (*Pinus contorta*) was introduced from North America and planted in Scotland from the late 1950s onward, large numbers of pine beauty moth larvae defoliated (removed the needles from) thousands of the newly introduced trees, causing them to die. Nearby Scots pine forests remained undamaged, because the natural enemies occurring there, such as parasitoid wasps (pp.68-69), kept moth numbers low.

BARK BEETLES Many species of beetle live within the wood and bark of trees, often remaining undetected until they have seriously damaged or killed the trees. Adult bark beetles, such as the eight-toothed European spruce bark beetle (*Ips typographus*), lay their eggs in a tunnel in the bark. When the larvae emerge, they tunnel into the wood of the tree, forming galleries. Most bark beetles live in dead or dying trees, but a few, such as the mountain pine beetle (*Dendroctonus ponderosae*), attack healthy trees. Some bark beetles also transmit disease, such as those that carry the fungus responsible for Dutch elm disease.

TENT CATERPILLARS Tent moth caterpillars are named for the silk nests they build, and they damage trees through defoliation. They are also called processionary moths because the caterpillars form lines, or processions, on trees and on the ground when they emerge from their tent. Examples include the pine processionary moth (*Thaumetopoea pityocampa*) and the oak processionary moth (*Thaumetopoea processionea*). These caterpillars can also have adverse health impacts for humans, livestock, and other animals. The hairs shed by older larvae can cause skin irritation and, in some cases, breathing difficulties.

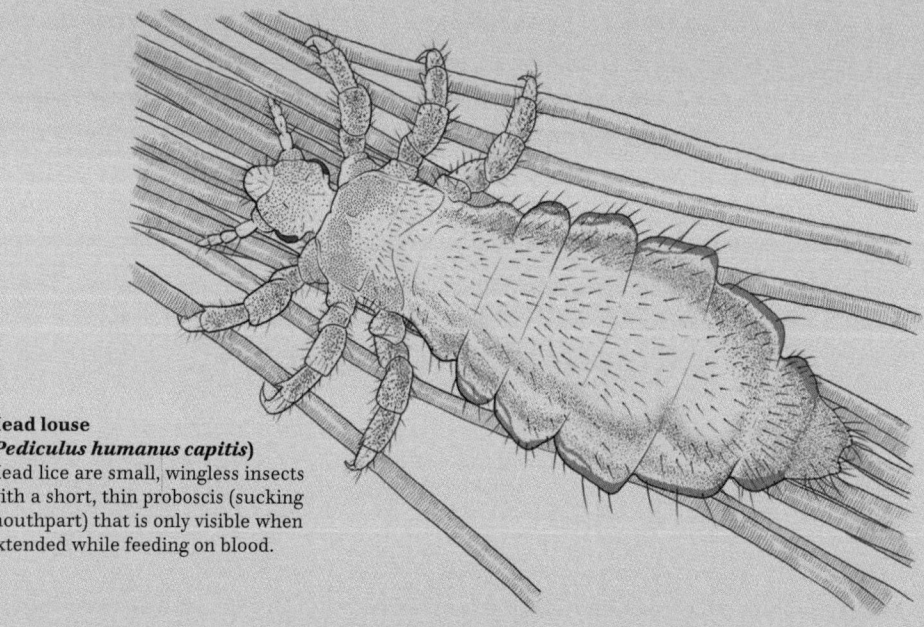

**Head louse
(*Pediculus humanus capitis*)**
Head lice are small, wingless insects with a short, thin proboscis (sucking mouthpart) that is only visible when extended while feeding on blood.

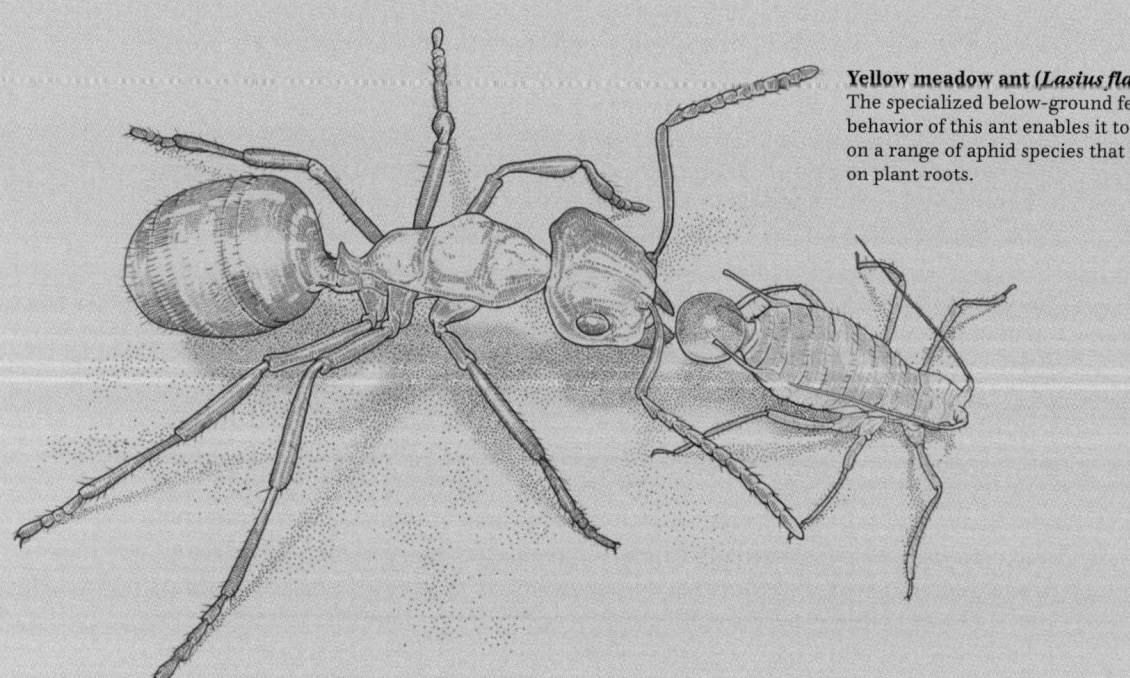

Yellow meadow ant (*Lasius flavus*)
The specialized below-ground feeding behavior of this ant enables it to live on a range of aphid species that feed on plant roots.

With an estimated 10 quintillion (1 billion billion) individual insects in the world, there are many specialist species—eating a limited diet or occupying a specific environment—that have adapted to coexist with other organisms.

SPECIALISTS

PARASITIC SPECIES The mere mention of the word "parasite" often conjures up feelings of disgust, but parasites are fascinating insects in that they have developed adaptations to allow them to succeed in their specialist environments. For example, both fleas and head lice (*Pediculus humanus capitis*) live on the skin of mammals and feed on their blood, but they have evolved different adaptations in order to do this. Loathed by schoolchildren and parents everywhere, the humble head louse lives its entire life cycle on the human scalp and transfers from host to host by close contact. Lice are so specialized to their environment of human hair that they cannot walk well on a flat surface. However, they are agile when moving from hair to hair: their legs are short, and the final leg segment is formed into a set of claws, which are adept at grasping hair. In fact, head lice are able to move quickly up and down shafts of hair, which means they can travel easily between hosts. In comparison, fleas only spend part of their life cycle on their host. The larvae live in the host's home (nest or other structure), feeding on any available organic material, whereas the adults live in the host's fur, feeding on blood. Adult fleas are far more able to travel between hosts, using their powerful back legs to jump to the next host with ease.

NONPARASITIC SPECIES Bees, ants, and wasps are specialists in how they interact with other plant and insect species. For example, where several species of solitary bees have evolved to collect pollen from specific plants, their life cycle may be tied to the annual patterns of those flowers. Long-horned bees in the genus *Melissodes* specialize on thistles, for example. Other species, such as flies in the genus *Prosoeca*, have developed mouthparts to access nectar-producing parts of flowers that other insects cannot reach. Some ant species, including the yellow meadow ant (*Lasius flavus*), have evolved to live below the ground alongside root-feeding aphids, which they milk for honeydew (a sugary solution).

Generalists are the most adaptable of the insects, able to take advantage of a wide range of resources and environments. They make a huge impact on their ecosystems and are often the first to exploit new opportunities.

GENERALISTS

GENERALIST POLLINATORS Tomato flowers, poppies, and lavender are all very different, and yet generalist pollinators such as bumblebees, honeybees, social wasps, butterflies, many beetles, and flies visit them all to feed on nectar and collect pollen. Among them, buff-tailed bumblebees (*Bombus terrestris*) are quite impressive. They are able to use techniques such as buzz pollination, whereby they produce strong vibrations directed at the anthers of the flower to cause the release of pollen. However, generalist pollinators cause more cross-pollination (when pollen from one plant is received by a different type of plant) than specialists and are, therefore, less efficient when pollinating some plants.

HUMAN INTERACTIONS Generalist insects are among the most successful invasive alien insects, able to adapt and survive in new environments after human introduction. Generalists such as silverfish, fruit flies, and cockroaches are also among the most likely insects to live in our homes. German cockroaches (*Blattella germanica*) have been living with humans for so long, their evolutionary history before this is not known! They may not always be appreciated, but generalist scavengers like cockroaches play very important roles in the wild, where they recycle waste and are essential to the health of ecosystems and the environment.

PLANT EATERS Generalist plant eaters can survive on a wider range of plants than specialists, but this means they risk choosing a plant that is not as good as some alternatives. The black bean aphid (*Aphis fabae*) switches between being a generalist and a specialist. During the summer months, winged aphids are generalists and they feed on several plants. When the fall comes, they migrate back to the primary host plant as specialists, feeding on spindle trees (*Euonymus europaeus*)—possibly a more stable lifestyle through the adverse winter months.

Black bean aphid (*Aphis fabae*)
Sap-sucking black bean aphids
can reach very high numbers on
their host plants. Honeydew (a sugary
solution) is excreted from the tiny
upright tubes called siphunculi
at the aphid's rear end.

**Rose chafer
(*Cetonia aurata*)**
The rose chafer beetle
plays an important role
in transferring pollen from
one flower to another
by collecting it from the
stamen and transferring
it to the stigma.

While many people think of bees as important pollinators, pollinating insects are incredibly diverse and span many different insect orders. Some are highly specialized, but many are generalists, visiting various types of flowers.

POLLINATORS

WHAT IS A POLLINATOR? Pollinators are not a defined taxonomic group. A pollinator is any animal that visits a flower for food (or to collect fragrant oils, or just to rest) and inadvertently picks up and transfers pollen to the female reproductive structures of the same or another flower. Nearly 90 percent of all plant species benefit from pollinators, including many human food crops.

WHO ARE POLLINATORS? The best-known pollinators are bees, who visit flowers to fuel every part of their life cycle: nectar gives adults energy to fly, and protein-packed pollen provides nutrition for larvae. Within the bees, the European honeybee (*Apis mellifera*) gets the most attention, but there are around 20,000 species of bee worldwide, most of whom do not live in colonies and do not produce honey. There are solitary bees that specialize in collecting fragrances from orchids (which they release in elaborate courtship displays), tiny stingless bees that form large colonies in trees in the tropics, and massive resin-collecting bees that live in termite nests. And beyond bees—flies, wasps, beetles, butterflies, bugs, and even crickets also act as important pollinators.

WHY IS POLLINATOR DIVERSITY IMPORTANT? Each species of pollinator varies in several ways and has adaptations (including body shape and size and even tongue length) to assist them in accessing resources from the flowers they visit. They have different ways of approaching and handling flowers, favored habitats and climates, and preferred seasons and times of day when they are active. This diversity is important for the effective pollination of plants across the world. Even in a single ecosystem, flowers can be visited by several pollinator species, and pollinators can visit several plant species. The interactions of different pollinating insects and flowers can be mapped to determine how resilient ecosystems are to changing climates and land use, disease, and pollution.

In both their larval and adult stages of life, insects play a crucial role as a source of food for many other species.

FOOD FOR ANIMALS

The biomass of insects worldwide is huge, with recent estimates of the global annual total of terrestrial arthropod prey exceeding 550 tons (500 metric tons). This colossal biomass represents an enormous source of food energy, and insects are, therefore, a vital part of food chains and food webs—which can be used to describe how each organism feeds on another within the same ecosystem. Food chains are diagrams that show how the energy flows from one species to another, and food webs consist of interconnected food chains. As such, food webs are a visual representation of the relationships among species within a community, and they illustrate the transfer of food energy from plants to herbivores to carnivores. They are also an important tool for understanding the interactions among predators and their prey.

MOSQUITOES AS FOOD FOR MANY ANIMALS Adult and larval forms of mosquitoes live in different habitats, with the adults being terrestrial flying insects and the larvae being aquatic. This feature of their biology helps reduce the competition for food between adults and their offspring. It also means that they occupy two different ecological niches within their ecosystem, making them important in more than one food web. In the food web opposite, mosquito larvae act as decomposers, eating plant debris that falls into the water they inhabit. The larvae themselves are a food source for small fish; aquatic insects; and even the larvae of other insects, such as dragonflies. Adult mosquitoes are consumed by adult dragonflies and also make a tasty treat for small lizards, such as the common lizard (*Zootoca vivipara*), and bats. Mosquitoes of the genus *Culex* have a special relationship with small perching birds, such as the barn swallow (*Hirundo rustica*), in that the adult mosquitoes feed on these birds but also get eaten by them. Lizards, swallows, and dragonflies are, in turn, eaten by even larger animals, including the common kestrel (*Falco tinnunculus*).

Food web (*Culex* spp.)
This food web shows how important mosquitoes can
be to their ecosystems. The larvae act as decomposers,
taking carbon from plant life and making the energy
available to animals further along the web.

Key

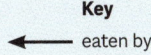
eaten by

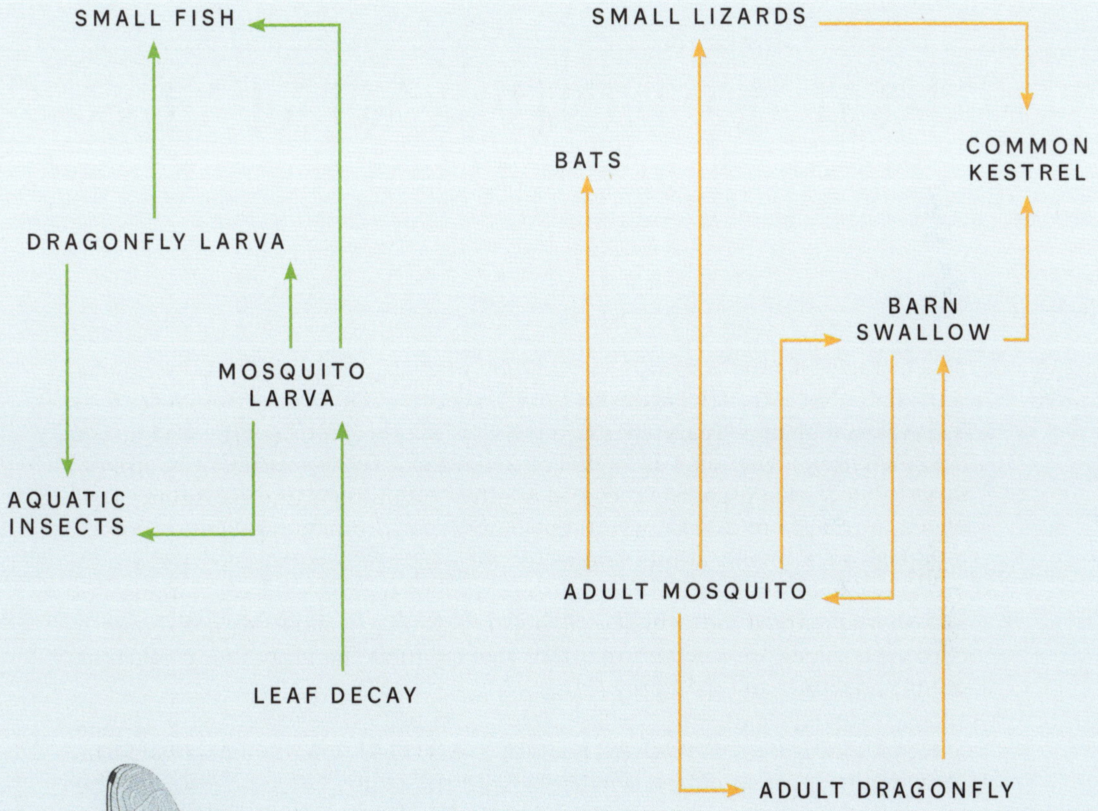

SMALL FISH

SMALL LIZARDS

BATS

COMMON
KESTREL

DRAGONFLY LARVA

MOSQUITO
LARVA

BARN
SWALLOW

AQUATIC
INSECTS

ADULT MOSQUITO

LEAF DECAY

ADULT DRAGONFLY

GOLDEN-RINGED
DRAGONFLY
(*CORDULEGASTER
BOLTONII*)

Some parasitic insects simply use their animal hosts for food and shelter and can be a minor irritation, while others spread major and sometimes fatal diseases.

ANIMAL PARASITES

A variety of insects live on animals as parasites and feed on their blood. In some species, such as louse flies or keds, both the males and females feed in this way. Within the many different species of louse fly, some feed only on birds and are adapted to move ably around within their feathers. For example, the pigeon fly (*Pseudolynchia canariensis*) is a parasite of domestic pigeons throughout the tropics and subtropics. Other insects, including the sheep ked (*Melophagus ovinus*), feed on mammals and cause both skin and wool damage. In blood-feeding insects, such as biting midges and mosquitoes, only the females bite and feed on blood because they require protein to produce eggs. Males feed on plant nectar secreted by extrafloral nectaries (plant glands).

FINDING A MEAL There are some biting midges that feed exclusively on frogs, and like their mammal-biting and bird-biting counterparts, they are very tiny—only 1–2 mm long. Frog-biting midges locate their host via the calls that the frogs make to each other, and some midges are only attracted to certain species. Other types of biting midges (for example, those in the genus *Culicoides*) find their mammalian host by sensing the chemical cues it emits, such as carbon dioxide and heat. This is also true of larger flies, including stable flies and deer flies, which bite animals and can be quite a nuisance.

GETTING STUCK IN Some parasitic insects have a more complex relationship with their animal host—for example, botflies. These flies have a larval stage (maggot) that develops within the skin, causing lesions, or within the gut, as commonly seen in horses. Some botflies even lay their eggs on another intermediate insect vector, such as a mosquito, so the eggs can be delivered to a suitable mammal host. Bee-grabbers (*Myopa* spp.) go one step further. They snatch a bee out of flight and inject their egg into the bee's abdomen, where it develops and eventually kills the host when pupation occurs.

DELIVERING DISEASE Many parasitic insects can spread disease-causing pathogens, such as viruses, bacteria, and protozoa. They take these pathogens into their body by feeding on the blood of an infected animal, and deposit them when they feed on a subsequent animal. Biting midges of the genus *Culicoides* spread bluetongue virus, which causes disease in ruminants such as sheep and cattle, and African horse sickness virus, one of the most lethal diseases for horses.

Specialist midges
Biting midges are found almost everywhere in the world, from farmland in the UK to the rainforests of Costa Rica, where some are specialized to feed on frogs. There are currently 120 described extant species of frog-biting midges.

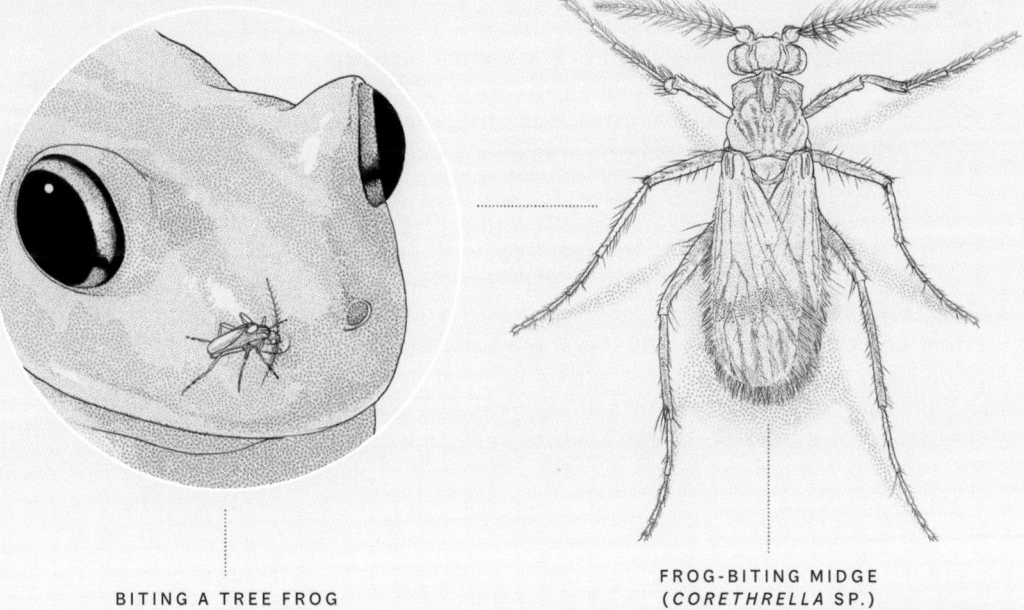

BITING A TREE FROG

FROG-BITING MIDGE
(*CORETHRELLA* SP.)

The overuse of antibiotics has led to the emergence of bacteria that are resistant to almost any antibiotic therapy and a worldwide rise in patients with antibiotic-resistant wound infections. There is now renewed interest in an ancient treatment for infected wounds: maggots.

MEDICINAL MAGGOTS

Maggot therapy essentially involves giving a patient an artificially induced infection from a fly, termed myiasis. Clinicians apply larvae (maggots) of the green bottle fly (*Lucilia sericata*) to a chronic infected wound. The larvae feast on the dead tissue, and by doing so eliminate infection. But these maggots are not any old wild maggots. Those used clinically are known as medicinal maggots, and specialist laboratories worldwide are licensed to produce them. These amazing creatures are reared under strict aseptic conditions, carefully packaged, then shipped. They can be supplied directly to hospitals and clinical centers for the treatment of all sorts of wounds, including leg ulcers and pressure sores, as well as infected surgical wounds, burns, and trauma injuries to humans.

HOW DOES MAGGOT THERAPY WORK? Trained clinicians place a number of tiny larvae onto a wound. Over a few days, the larvae not only clear away the dead wound tissue, but also eliminate infection. It has been shown that maggots combat wounds in three different ways: debridement (getting rid of dead tissue), disinfection (getting rid of bacterial infection), and healing (accelerating the healing rate). The advances in our understanding of maggot therapy and its treatment stem from both clinical reports and the results of laboratory investigations.

WHY THESE MAGGOTS? Green bottle fly larvae are able to achieve speedy and effective wound debridement because of their ability to rapidly consume and ingest dead tissue. When it comes to feeding, green bottle fly larvae need to

acquire resources as quickly as possible because slow or prolonged feeding may increase their exposure to predation in the wild. They have, therefore, developed particularly effective mechanisms to rapidly acquire all the resources necessary for growth. Their saliva contains potent enzymes that quickly break down and digest tissue. These secretions turn dead wound tissue into a sort of soup, which they then slurp up.

SAVING LIMBS Just a few days of maggot therapy can save an infected limb from amputation. In the past decade, thousands of patients have had their wounds successfully treated with maggots. These insects are nature at its very best, and for chronic wounds, maggots are doing something very amazing.

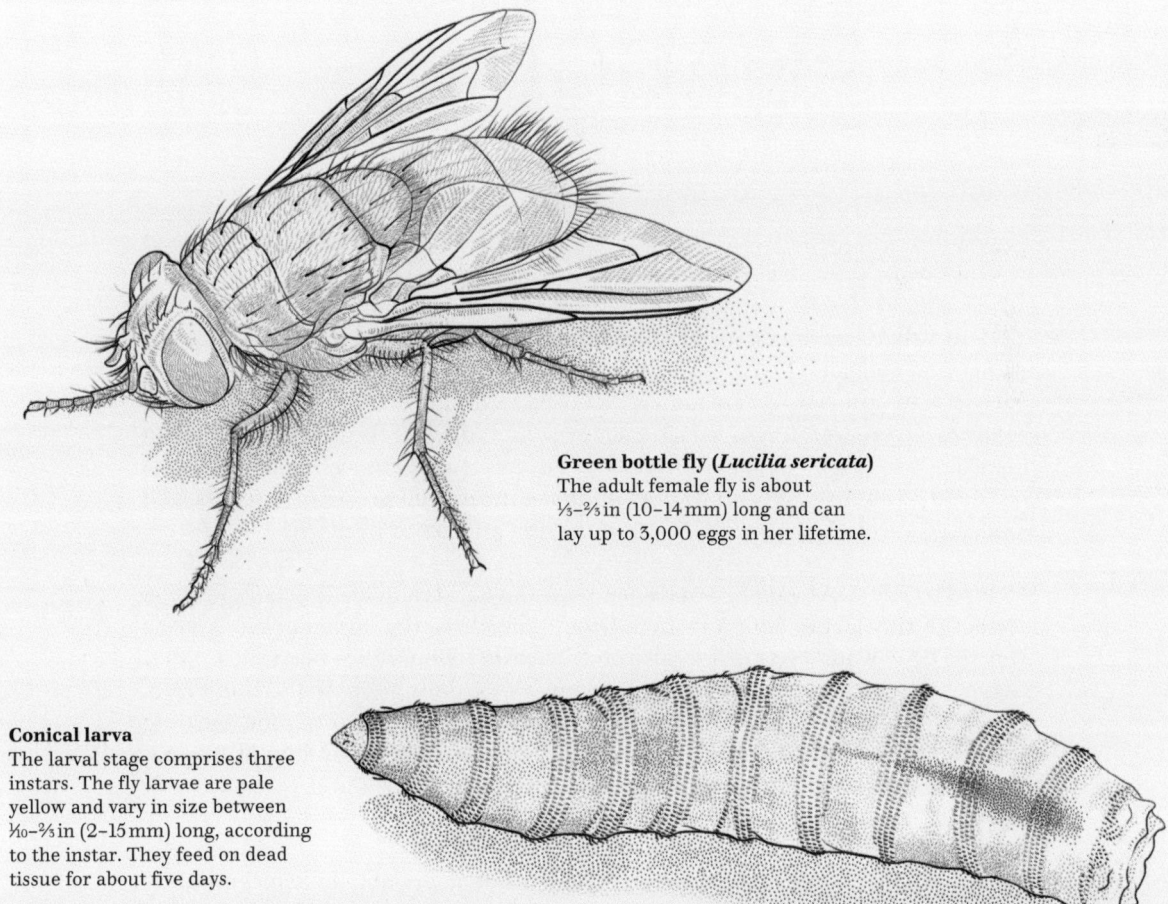

Green bottle fly (*Lucilia sericata*)
The adult female fly is about ⅓–⅔ in (10–14 mm) long and can lay up to 3,000 eggs in her lifetime.

Conical larva
The larval stage comprises three instars. The fly larvae are pale yellow and vary in size between ⅒–⅔ in (2–15 mm) long, according to the instar. They feed on dead tissue for about five days.

Common wasp (*Vespula vulgaris*)
Also known as yellowjackets, wasps
in the genus *Vespula* are abundant in
the northern hemisphere. The common
wasp and the German wasp (*Vespula
germanica*) are very similar in appearance,
the latter being slightly larger.

Tropical bed bug (*Cimex hemipterus*)
Bed bugs such as this species and
the closely related *Cimex lectularius*,
which lives in more temperate regions,
feed on human blood at night by
piercing the skin with their sharp
mouthparts. Similar to other parasitic
insects that do not invade the body, bed
bugs do not have functional wings.

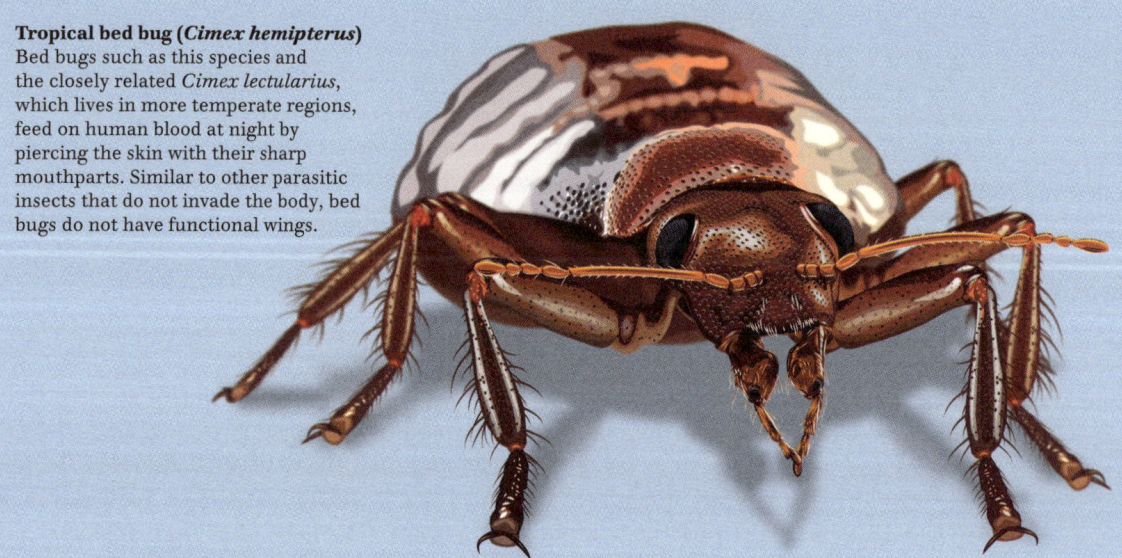

Insects that bite or sting people are commonly regarded as a nuisance, but some of them play important roles within our ecosystems as pollinators or pest controllers.

INSECTS THAT BITE AND STING

STINGERS Of the approximately 100,000 known wasp species, only about 33,000 species are hunting wasps and carry stings. Social wasps sting in defense when threatened, delivering venom into the skin through a modified ovipositor. Solitary wasps either sting to defend themselves or to immobilize their prey. All wasps are very valuable within our ecosystems. Common wasps (*Vespula vulgaris*), for example, are predators and help control crop pests such as aphids, and fig wasps (which are stingless) are known to pollinate more than 750 species of fig tree. Bees, including bumblebees, also play a crucial role in pollinating crops and, therefore, in global food supplies. They are highly effective at collecting pollen. Hairs on their body attract pollen grains, which are then collected on their legs or transported on their abdomen to other plants or back to the nest to feed developing larvae.

BITERS Some biting insects, such as mosquitoes and midges, seek out animals and humans for the sole purpose of feeding on their blood, but do not live on them. Other biting species, such as lice (p.159), spend most or all of their life on humans. Lice are wingless insects, and some species specifically live in human body hair and have claws that help them hold onto individual hairs. In fact, they cannot survive away from their hosts beyond about 24 hours. The immature nymphs and adults of both sexes feed on blood around three times a day. Bed bugs also feed on human blood, and at around ¼ in (6 mm) in length are about twice as large as lice. Unlike lice, though, bed bugs do not spend all of their time on the host. They live in crevices of furnishings within the home, where they can easily access a nightly meal from a human and lay their eggs.

While insects bring many benefits to both the environment and society, the transmission of infectious microorganisms by insects and other arthropods is responsible for 700,000 human deaths per year.

HOW DO INSECTS SPREAD DISEASE?

Most microorganisms are harmless or even beneficial to humans, but there is a small group of infectious microorganisms (pathogens) that can cause disease. An insect vector is an insect that can transmit an infectious microorganism between animals, including humans. There are two types of insect vectors: mechanical and biological.

MECHANICAL VECTORS Insects acting as mechanical vectors move microorganisms from one surface to another—the pathogens do not develop or reproduce inside the insect. The relationship between these insects and the microorganisms is simple and sometimes coincidental. Pathogens transmitted mechanically are usually picked up by the insect on its mouthparts or feet when it feeds in unsanitary conditions. The insect then moves the microorganisms directly to the human host, for example, while feeding on eye secretions, or indirectly by contaminating food.

BIOLOGICAL VECTORS An insect acting as a biological vector does not just move a microorganism from one place to another: it is a crucial part of the life cycle of the pathogen. Biological vectors, such as mosquitoes, usually ingest microorganisms from the blood of a human or other animal. The microorganisms then pass through the insect's digestive system, sometimes developing until they become infectious and often reproducing until they are present in great numbers. Once infectious, the microorganisms are usually found in the saliva or excrement of the insect. The insect will then transmit the microorganisms to a human when it salivates, urinates, or defecates during or after feeding on their blood.

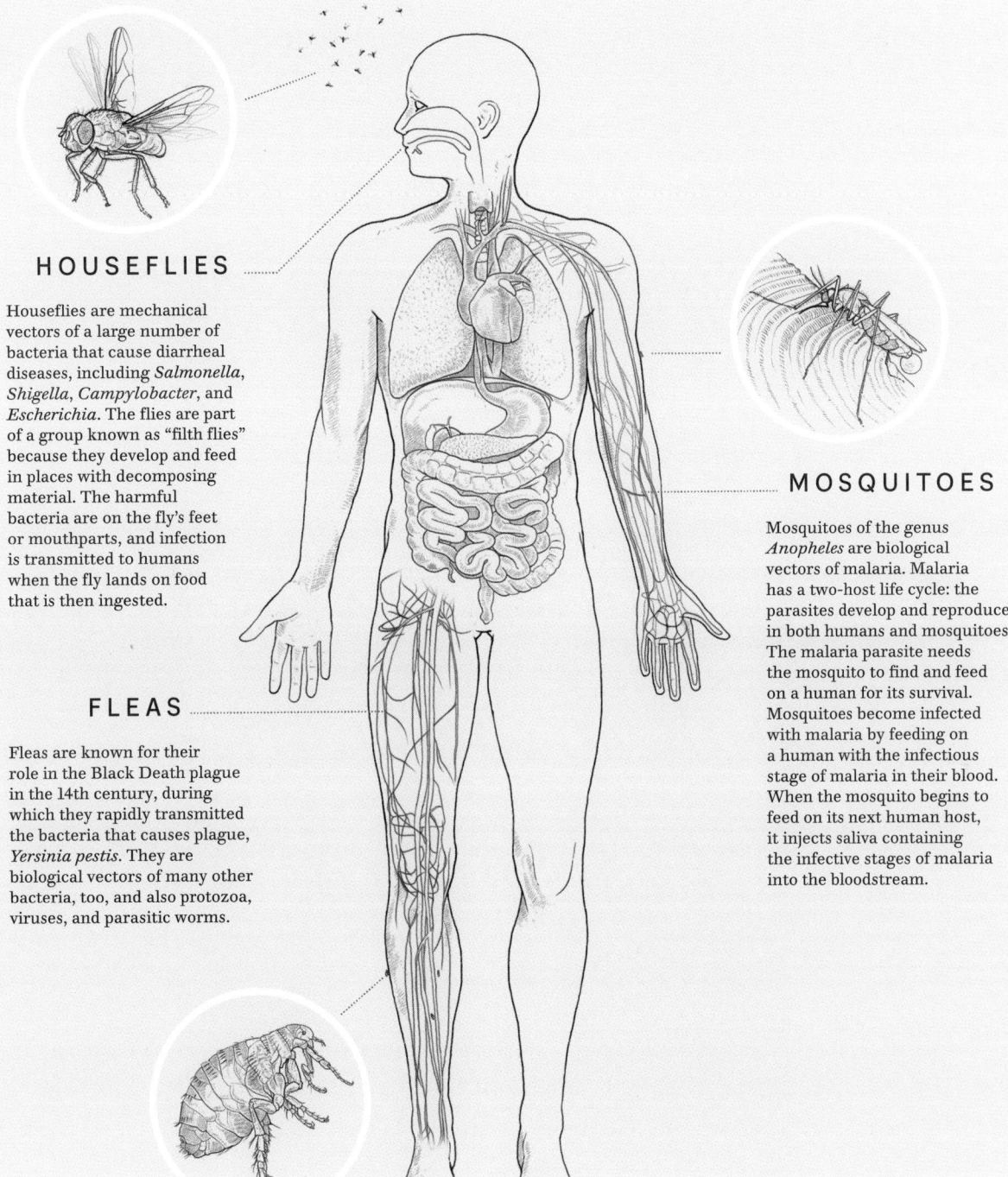

HOUSEFLIES

Houseflies are mechanical vectors of a large number of bacteria that cause diarrheal diseases, including *Salmonella*, *Shigella*, *Campylobacter*, and *Escherichia*. The flies are part of a group known as "filth flies" because they develop and feed in places with decomposing material. The harmful bacteria are on the fly's feet or mouthparts, and infection is transmitted to humans when the fly lands on food that is then ingested.

FLEAS

Fleas are known for their role in the Black Death plague in the 14th century, during which they rapidly transmitted the bacteria that causes plague, *Yersinia pestis*. They are biological vectors of many other bacteria, too, and also protozoa, viruses, and parasitic worms.

MOSQUITOES

Mosquitoes of the genus *Anopheles* are biological vectors of malaria. Malaria has a two-host life cycle: the parasites develop and reproduce in both humans and mosquitoes. The malaria parasite needs the mosquito to find and feed on a human for its survival. Mosquitoes become infected with malaria by feeding on a human with the infectious stage of malaria in their blood. When the mosquito begins to feed on its next human host, it injects saliva containing the infective stages of malaria into the bloodstream.

Bugs *are not going to inherit the earth. They* **own it now.** *So we might as well* **make peace** *with the landlord.*

~ Thomas Eisner, 1989 ~

Our homes provide shelter and in some cases food for many insects. From the lacewings and ladybugs that spend winters in window frames to the tiny insects that feed on clothes and carpets, some are more welcome than others.

INSECTS IN THE HOME

Most people enjoy the occasional sighting of a butterfly or ladybug but consider other insects in the home to be a nuisance and even harmful. Some insects feed on our belongings, such as soft furnishings and food, and others such as cockroaches and mosquitoes transmit diseases. It is, however, fascinating to consider the abilities of these so-called synanthropic species that live in close association with us and benefit from our ways of living. Clothes moths and carpet beetles, in particular, are unusual in their adaptation to feed on keratin.

ON CLOTHING AND FURNISHINGS One of the most familiar insects in the home is the common clothes moth (*Tineola bisselliella*). Its destructive larvae feed on natural fibers—including wool, silk, and cotton—and can chew holes through garments and soft furnishings, leaving behind a trail of damage. As their name suggests, carpet beetles (for example, *Anthrenus verbasci*) feed on carpets, but they also consume clothing, upholstery, and even stored food items. Carpet beetle larvae cause significant damage by feeding on wool, fur, and feathers.

IN THE PANTRY Species of grain beetle vary slightly in appearance and behavior, but they all share a preference for eating food products that humans also enjoy, including grains, cereals, nuts, and spices. Grain beetle larvae consume the insides of grains, but also contaminate food with their excrement, cast-off skin, and silk webbing, reducing both the nutritional value and quality of the grain. In some cases, the food is rendered unfit for human consumption. Common species include the sawtoothed grain beetle (*Oryzaephilus surinamensis*) and the merchant grain beetle (*Oryzaephilus mercator*).

Dung beetle (*Trypocopris pyrenaeus*)

DISCOVER

CHAPTER IV

INSECTS ARE ALL AROUND
us in everyday life, but these tiny creatures
can also survive and thrive in some of the most
extreme environments on Earth. Explore these
different habitats to discover insects living in
forests, caves, deserts, and puddles; up
mountains; and deep within the soil. Go outside
for a walk and start to notice the insects near you.

Insects have lived on Earth *far longer than humans have ... and they've found at least **a million different ways** to make a living here; we're living on **their planet**, not the other way around.*

~ May Berenbaum, 2011 ~

Over time, insects have evolved to live in a wide range of environments on every continent on Earth, from the dizzying heights of tropical rainforests to the decidedly chilly ponds of melted Arctic ice.

HABITATS

The ancestors of modern insects evolved on land around 480 million years ago. They diversified to live in every available land habitat, from desert and tundra, to woodland and grassland. As well as their success on land, more than 6 percent of insects, including immature stages of caddisflies (p.173), mayflies (p.147), and stoneflies (p.149), spend some of their life cycle in freshwater.

DIVERSITY There are common patterns in the distributions of insects across the world. Insects are cold-blooded, so they are sensitive to temperature and moisture. They can occur in extremely hot (pp.134–135) and extremely cold places (pp.136–137). However, their diversity is highest in hot, humid environments, such as tropical regions near the equator, and lowest in cold areas, such as polar regions and mountaintops. Insects have evolved alongside flowering plants, and around half of insects feed on plants. It is no surprise, therefore, that global patterns of diversity and abundance in insects follow those in plants.

PATTERNS OF CHANGE The numbers of insects and where they live have seen substantial changes in recent years, mainly because of human activities. Rising global temperatures are causing many insects to shift their ranges (pp.200–201), which has led to interactions among species being disrupted. For example, mismatches in timing can occur between insects and the availability of their food plants. Changes caused by agriculture mean that many insect habitats are being lost, unless farms have flower-rich field margins and hedgerows to help make space for nature. Global trade and transporting goods and people have led to the accidental introduction of insects to new regions, where they may disrupt local ecosystems (pp.198–199). Understanding the factors that cause changes in insect habitats is crucial for conserving insect biodiversity.

TERRESTRIAL HABITATS

Numbering an estimated 5.5 million species, insects are the unsung heroes of life on land, dwelling in an astonishing array of habitats. The term "terrestrial" refers to everything except marine and freshwater environments.

FORESTS For insects, forests are treasure troves of biodiversity. Insects thrive in the soil, leaf litter, and underlying layer of vegetation beneath trees. They also hide under tree bark, reside in hollow stems and twigs, or live on the surfaces of leaves and branches. Some insects, such as ants (pp.48–49), have evolved close relationships with plants, which produce shelters (domatia) within their leaves, stems, or thorns for ants to live in. Other insects induce the formation of plant galls (abnormal growths), which provide food and shelter. Many dead-wood-loving insects (pp.110–111) make their homes in fallen trees and decaying logs.

FOG-BASKING BEETLE
(*ONYMACRIS UNGUICULARIS*)

HAWKMOTH PUPA
(SPHINGIDAE)

GAVARNIE BLUE BUTTERFLY
(*AGRIADES PYRENAICUS*)

HAWAIIAN CAVE PLANTHOPPER
(*OLIARUS POLYPHEMUS*)

GRASSLANDS AND BEYOND With their grasses and wildflowers, grasslands offer many resources for insects to exploit. By contrast, arid hot and cold deserts and icy tundras are the ultimate testing grounds for insect survival, and species here have evolved extraordinary adaptations to endure the harshest of conditions. Examples of insects surviving in other extreme environments include pale and blind cave-dwelling species (pp.138–139) and high-elevation butterflies and flies that live at sites up to 19,685 ft (6,000 m) above sea level (pp.128–129). Even among the concrete and steel of bustling cities, insects can survive and live alongside people (pp.130–131).

ABOVE AND BELOW GROUND Subterranean insects live and forage in soil and sand, while leaf-litter-dwelling insects exist at the margins of above and below-ground environments. Ants, for example, create intricate nests for their complex societies, while other solitary insects exploit the spaces between soil, rocks, and roots. Around 400 million years ago, insects took to the skies, evolving flight (pp.42–43). Numerous insects can be found tens of thousands of feet above the ground, using air currents to move around but often ending up as food for birds.

Monarch butterflies (*Danaus plexippus*) These butterflies migrate over long distances from their breeding grounds in North America and Canada to their overwintering sites in the temperate forests of Mexico and California.

Alpine longhorn beetle (*Rosalia alpina*) This large beetle is most common in temperate beech forests with older, dead, or dying trees. Management practices that involve removing such trees are, therefore, damaging to this species, but "dead wood islands" have been proposed as a conservation measure.

Temperate forests make up about a quarter of the world's forests and provide numerous places for insects to thrive, from leaf litter and undergrowth to leafy treetops. They can support many different insect species, but climate change is bringing new challenges.

TEMPERATE FORESTS

HOME TO MANY All temperate forests share distinct layers of vegetation. Butterflies, moths, and beetles are found in the canopy; bees, wasps, and bugs visit branches, fruits, and flowers in the understory layer; and ground-dwelling beetles roam the leaf litter on the forest floor. The many insect species vary according to whether the forest contains deciduous broad-leaved trees (such as oak, maple, and beech) or coniferous trees (including pine, fir, and spruce).

THREATS TO INSECTS The weather in temperate forests changes with the seasons, exposing insects to both cold and warm conditions. In such habitats, insects have evolved survival strategies, such as overwintering until conditions improve or migrating to warmer locations. Temperate forests are also affected by human activities, including cutting down trees, planting non-native trees, and introducing tree pests and diseases. In fact, several insect species are threatened by extinction due to extensive logging, pollution, and deforestation. One example is the Alpine longhorn beetle (*Rosalia alpina*), which occurs in beech forests but is threatened by loss of habitat and so is protected in many European countries. Over the past decades, insects in temperate forests have faced multiple challenges as a result of the trees they depend on becoming stressed from droughts and warming temperatures that can cause forest fires. These climate changes can also trigger insect pest outbreaks. For example, a recent outbreak of the mountain pine beetle (*Dendroctonus ponderosae*, pp.80–81), which introduces a fungus when it lays its eggs under tree bark, has caused extensive damage to the lodgepole pine trees of western North America, losing carbon stored in the forests and reducing the forest's capacity to take up more carbon from the atmosphere.

Tropical forests contain vast numbers of insects and more species than any other habitat on Earth. High temperatures and rainfall produce conditions in which plants and trees thrive, providing an amazing diversity of food and habitats for insects.

TROPICAL FORESTS

EXCEPTIONAL DIVERSITY Most insects in the world occur in tropical forest habitats, and most species that live in tropical forest habitats are insects. Studies of a site in Panama estimate that for every species of plant (1,294), there are likely to be at least 17 arthropod species. This exceptional diversity is one reason why deforestation is such a huge concern. In terms of color and size, tropical forests contain a large array of insects. For example, Rajah Brooke's birdwings (*Trogonoptera brookiana*) of Southeast Asia have vivid metallic green wing markings and a wingspan of 6–7 in (15–17 cm). Many species of lanternfly (so named because their very long snout was thought, at one time, to be luminous) are also very colorful and have distinctive patterning. The wings of *Kallima* butterflies are brightly colored when open but various shades of brown when closed, so the butterflies resemble dead leaves when skulking in the leaf litter.

DIFFERENT HABITATS Many insects, such as leaf-litter ants, live in the dense understory of forests, where light levels are low, humidity is high, and temperatures are buffered from the hot daytime conditions that are experienced in the canopy. When trees die and fall over, they form gaps in the forest that can also be hotspots for insects, such as those that feed on dead wood (pp.110–111). Furthermore, trees in tropical forests support other plants such as bird's nest ferns (*Asplenium nidus*). These large ferns (up to 441 lb/200 kg in weight) grow on tree trunks and branches and trap falling leaves. A single plant can house as many insects as the entire tree it grows on. Although there are threats to tropical insects when forests are converted to grow crops, such as oil palm, some insects thrive in these plantations, including palmking butterflies (*Amathusia phidippus*) and Asian rhinoceros beetles (*Orytes rhinoceros*), which eat young palm trees.

EMERGENT LAYER

CANOPY

UNDERSTORY

FOREST FLOOR

MALE RAJAH BROOKE'S
BIRDWING (*TROGONOPTERA
BROOKIANA*)

LANTERNFLY
(*PYROPS CANDELARIA*)

Dark and humid to hot and sunny
The wide variety of trees and plants
that grow in tropical forests produce
an amazingly complex structure,
which provides many places for
insects to live.

Dead wood from the decaying remnants of once-towering trees is an important habitat that supports a wide range of insects. Those living in this woody world use the moist, stable environment for food and shelter.

DEAD WOOD

RECYCLING

Wood-boring insects, including the iridescent flatheaded or metallic wood-boring beetles, are important decomposers. These beetles use their strong mouthparts to dig tunnels for laying their eggs. The larvae, sometimes called woodworms, feed on the decaying wood, breaking it down and providing access to the next wave of decomposers—comprising other insects, microbes, and fungi—which continue the decomposition process. In this way, the nutrients locked up in wood are recycled.

BEETLES Dead wood is rich in the various components of woody plants (cellulose and lignin), which makes it an ideal habitat for many different types of beetles. Some bore under the bark, such as tiny bark beetles and snout-nosed beetles, whereas others bore into the wood, including the notorious deathwatch beetle (*Xestobium rufovillosum*), known not only for its distinctive tapping sound to attract mates, but also for the damage it causes to buildings with timber structures. Wood-boring larvae produce fine boreholes that serve as shelter for other insects, such as rove beetles, who prey on them, and also woodwasps, horntail wasps, and carpenter bees.

TERMITES AND OTHER INSECTS Termites are the true architects of the dead wood realm. These highly social insects break down woody tissues with the help of symbiotic microbes in their gut. Much-maligned due to their destruction of wooden buildings, these insects possess a remarkable ability to not only digest wood, but also to modify its structure by creating tunnels and galleries. These tunnels are infiltrated by water and therefore promote the growth of fungi and cause further dead wood decay.

A diverse array of other insects (including ants), predatory invertebrates, and tiny mites are found in rotting trees and logs. Along with fungi and microbes, they make up a rich, interconnected dead wood community. It is important to conserve their vital habitats by leaving dead wood to decay naturally.

Longhorn beetle (*Rutpela maculata*)
Longhorn beetles benefit from diverse woodland, thriving on a combination of dead wood and flowering plants. The larvae feed on the wood, whereas adults feed on pollen and nectar.

European sculptured pine borer (*Chalcophora mariana*)
Also known as the flatheaded pine borer, this insect breeds mostly in logs, stressed or dead pines, and pine stumps. It lays eggs in cracks and crevices in the bark; the larvae develop under the bark, then tunnel deeper into heartwood, where they pupate after several years.

Soil is home to 95 percent of insects at some time
in their life. They make their homes in the air spaces
and water surfaces that surround soil particles.

SOIL

WHO LIVES HERE? There is a wide variety of predators, including insects such as beetles and invertebrates such as spiders and centipedes, that dwell in soil. These predators feed on small decomposer invertebrates (those that break down dead plants and animals to return nutrients to the soil), so they play an important role in regulating their populations and the health of the soil. Soil-dwelling insects that are decomposers and scavengers include woodlice and silverfish. They feed on decaying plant material, fungi, and bacteria, so they contribute to the recycling of soil nutrients

Insects such as ants and termites are important soil engineers. They alter the structure of the soil by digging and building "houses" for their colonies. This activity benefits the soil by helping air, water, and nutrients to circulate. In addition, many moths overwinter as pupae in soil, leatherjacket and chafer grubs feed on plant roots, adult butterflies drink water and salts from the soil surface, and antlion larvae dig pits in sandy soils to trap and feed on invertebrates that fall in.

NURTURE Reducing activities that damage soil by compacting it helps nurture soil insects. Farmers can limit plowing by using machinery to sow seeds directly, and gardeners can try "no dig" gardening. Using natural agents, such as ladybugs, for the biological control of pests instead of chemical pesticides (pp.74–75) reduces the risk of pollution and toxic run-off into soil and therefore helps maintain soil insect populations.

Hawkmoth pupa (Sphingidae)
Hawkmoth caterpillars burrow into soil near their food plant in fall, then pupate and overwinter there. They emerge in spring, when temperatures are warmer.

Springtail (*Megalothorax minimus*)
No longer classified as insects, springtails are particularly abundant in soil and are important decomposers. To move through soil, they navigate past the water surfaces and through the air spaces between particles of silt, sand, and clay.

**European beewolf
(*Philanthus triangulum*)**
A female beewolf takes a honeybee
(*Apis mellifera*) into its nest. Honeybees
are the main prey of the beewolf across
its world range, but it may take other
bee species, such as the yellow-legged
mining bee (*Andrena flavipes*) in the UK.

**Northern dune tiger beetle
(*Cicindela hybrida*)**
An adult northern dune tiger
beetle can reach a speed of
8 ft/s (2.5 m/s) but needs very
warm weather conditions to do so.

*Sandy habitats are challenging environments
for many insects, but some species are well-adapted with
behaviors that make them highly successful predators
within these hunting grounds.*

SAND

LIFE IN SAND Sandy habitats are home to a range of insects, such as the Great Sand Dunes tiger beetle (*Cicindela theatina*) in the US and the northern dune tiger beetle (*Cicindela hybrida*) in Europe—the latter identified as one of the fastest beetles on Earth. Solitary wasps and bees thrive in sandy soils because they find it an ideal environment to build their nest. They spend the summer months burrowing and tunneling through the loose soil, creating small cavities in which to lay their eggs. These tunnels can merge into a vast intricate mesh right beneath our feet. Eggs are deposited in underground nest chambers, along with food provisions—pollen for bees, insect prey for wasps—and adults later emerge and fly away to find mates.

HUNTING IN SAND Sandy environments make ideal hunting grounds, and insect predators take a variety of approaches. For example, the pantaloon bees (*Dasypoda hirtipes*) dig into the sand to look for food, pushing back the spoil with their bright yellow, flared hind legs and leaving small triangular patterns in their wake. Beewolves (*Philanthus* spp.) also dig into the sand to find prey, whereas dune robberflies (*Philonicus albiceps*) patrol the sand dunes for smaller flies and bees, which they ambush on flower heads or catch in flight. Satellite flies are small parasitic insects that scout around in the sand, sneaking into the nests of solitary bees and wasps to lay their eggs onto the larvae of their insect host. After hatching, satellite flies then consume their host, along with any food provisions in the nest. Some insect predators such as tiger beetles (*Cicindela* spp.) hide under the sand surface, where larvae lie in wait for small insects and invertebrates before swiftly ambushing them. In the most extreme sandy habitats—deserts—insects have to be especially resourceful, and they have developed particular behaviors and adaptations to survive in the heat (pp.134–135).

Fields and grasslands vary considerably according to the diversity of plant species within them and the physical traits of those plants, such as the size of their flowers and amount of nectar they produce. The insects that call them home are just as diverse.

FIELDS AND GRASSLANDS

GRASSLAND RESOURCES The species of plants that occur, the type of grassland (such as acid, chalk, or limestone), and the way in which the land is managed (for example, whether sheep or cattle graze there) all affect the resources (leaves, nectar, pollen, and so on) available to grassland insects. And because the lives of grassland plants and insects are closely linked through pollination and herbivory, if either plants or insects decline, parallel declines can be observed in the insect or plant partner.

INSECT DIVERSITY The many types of grassland habitats result in a high diversity of grassland insects, whose behavior is varied, even among close insect relations. For example, solitary bee species that occur in grasslands and collect pollen from different flower species (polyectic) vary in their nest behavior. The red mason bee (*Osmia bicornis*) uses clay to build its nest in a variety of different types of cavities, whereas the two-colored mason bee (*Osmia bicolor*) exclusively nests in the shell of only four species of snail. Within a small area of hogweed (*Heracleum sphondylium*), there can be hundreds of insects, ranging from wasps to beetles to butterflies. Some use mimicry (pp.34–35) to disguise themselves as other species. For example, the hoverfly *Volucella bombylans* mimics different species of bumblebee, and the hornet moth (*Sesia apiformis*) mimics its namesake to protect it from being eaten.

GROUND ACTIVITY

While some insects pollinate grassland plants, beetles, ants, and springtails move around on and below the soil surface. Many ground insects feed on dead animal and plant material. Common in Europe, the horned dung beetle (*Copris lunaris*) feeds on animal dung and breaks down organic matter, thus returning nutrients to the ecosystem. Other ground insects are predators, such as the green tiger beetle (*Cicindela campestris*), which hunts caterpillars, ants, and spiders.

POLLEN BEETLE
(*MELIGETHES* SP.)

MARMALADE HOVERFLY
(*EPISYRPHUS BALTEATUS*)

Hogweed (*Heracleum sphondylium*)
This common plant, native to Europe and Asia, is host to many insect species.

PARASITOID WASP
(*ICHNEUMON SARCITORIUS*)

7-SPOT LADYBUG
(*COCCINELLA SEPTEMPUNCTATA*)

Throughout the evolution of insects, several groups—such as dragonflies, mayflies, midges, blackflies, and stoneflies—have successfully colonized rivers and streams.

RIVERS AND STREAMS

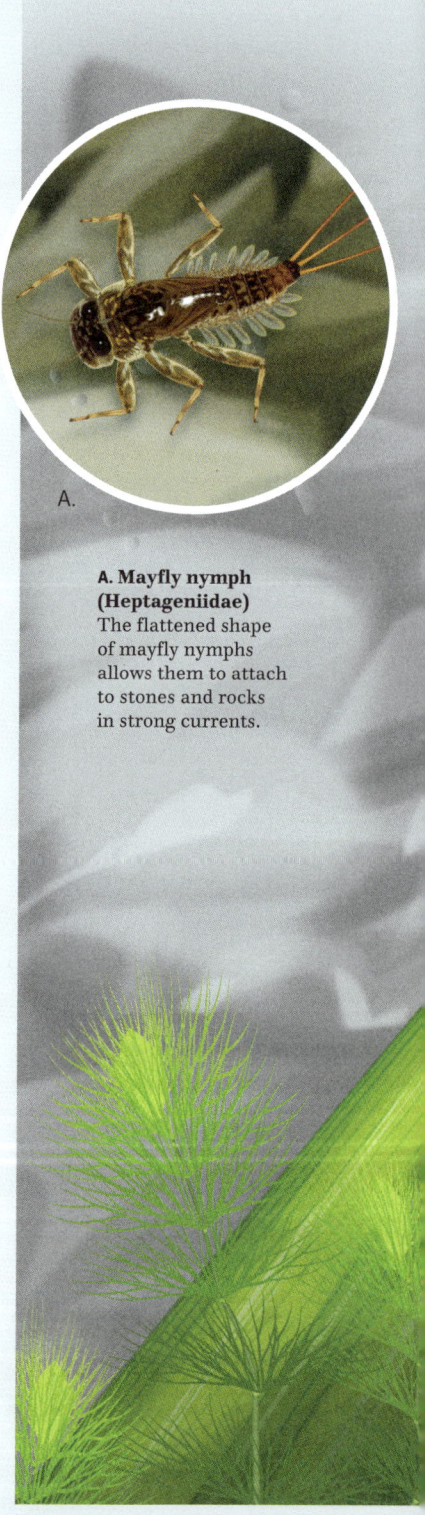

A.

A. Mayfly nymph (Heptageniidae) The flattened shape of mayfly nymphs allows them to attach to stones and rocks in strong currents.

Insects were well-suited to moving from land to rivers and streams, because during their evolution, they developed a waxy outer coating (exoskeleton) that made them waterproof and prevented them from drying out on land.

LIFE IN WATER Some insects that are adapted to spend part of their life cycle in freshwater have incomplete metamorphosis (pp.30–31), developing through several nymphal stages underwater before emerging to live on land as adults, then laying eggs in water. Mayfly nymphs scrape organic matter from stones and hard surfaces under the water, and the nymphs of some species have become flattened to make them streamlined. Damselfly nymphs capture prey in the water using their extendable mouthparts.

Insects that have complete metamorphosis (pp.28–29) have also evolved immature forms that are well-adapted to life in water. Some need shelter from fast-flowing currents, whereas others have adapted to exploit these. Some species produce silk to support their aquatic lifestyles; for example, caddisfly larvae spin nets from silk to catch food, and anchor themselves to the riverbed using hooks. The larvae of some species of midges spin silk into tubes and live within them, attached to the riverbed.

C. Stonefly nymph (Perlidae)
Stonefly nymphs feed on algae, plants, and invertebrates. They can be distinguished from mayfly nymphs because they usually have two "tails," while mayflies always have three.

C.

B. Blackfly larva (Simuliidae)
These larvae produce silk that is formed into pads to anchor them to the riverbed. They have intricate, fanlike structures on their head that capture very fine particles of food from the water that flows over them.

B.

D. Caddisfly larva (Hydropsychidae)
Most caddisfly larvae live in cases made of sand, small stones, or plant debris. The larvae in this family construct nets to catch food (pp.58–59).

D.

**Banded demoiselle
(*Calopteryx splendens*)**
This dragonfly occurs along
slow-moving rivers and
streams, ponds, and lakes
across most of Europe and
much of Asia.

Lakes often support high levels of insect biodiversity. Depth, temperature, and the chemical composition of the water greatly affect the numbers of insects that live there.

FRESHWATER: LAKES

LAKE HABITATS The type of insects found in lake habitats differs according to factors such as the rocks and soil underlying the lake, whether the landscape is flat or mountainous, and how isolated the lake is from other insect communities. How the nutrients flow through the lake and how it is drained will also shape the insect communities that live there. Some lakes have near constant water levels, whereas others may dry up for a time. These changing conditions have adverse effects on plants, which in turn affect the lake insects that feed on them.

LAKE COMMUNITIES Insects that commonly occur in lakes include true flies (pp.174–175), beetles (pp.170–171), true bugs (pp.160–161), and caddisflies (p.173). Some feed on detritus in the water or on small aquatic animals and plants; others may be predatory, hunting insect prey. Lake insects themselves are an important food source for other animals that make up the lake community, such as fish, frogs, and birds. Nonbiting midges can occur in freshwater lakes in their millions, and their immature stages feed on the plant and animal detritus lying on the lake bed. When the midges emerge as adults, these swarms of insects attract other aquatic insects, such as adult dragonflies, which feed on them.

CONSERVING LAKE INSECTS Lake margins are important habitats for insects. Here, the water is often warm and shallow, with many different types of aquatic and marginal plants. However, these areas can be disturbed by people and by livestock coming to drink, so conservation management is needed. In drier areas of the world, lakes are also used to irrigate crops, and loss of water and lower water levels can be very damaging for lake insects.

Ponds and puddles are important habitats for aquatic insects, but it can be challenging for insects to live in those that are prone to drying up or are far away from other wetlands and, therefore, difficult to find.

PONDS AND PUDDLES

LIFE IN STILL WATER Waterlilies float on pond surfaces, and pond margins are often flanked by reeds and rushes, all of which provide shelter and food for insects. Those that live in ponds are adapted to still water, which has a lower oxygen content than faster-flowing water in rivers and streams. Their adaptations include greater physiological tolerance to such conditions, but they are also adapted to migrate to a more suitable environment. Living in puddles is more difficult for insects, because puddles are only temporary and they regularly dry up.

GHOST PONDS Historically, the UK landscape had a vast number of ponds, created from digging to extract clays and other materials to use in marling (improving the soil structure and nutrient content) and for making bricks. This changed after the intensification of agriculture from the 1950s onward, and probably more than half a million ponds were lost because they were filled in to provide more land to grow crops. People are now searching for these "ghost" ponds and resurrecting them, creating more habitats for aquatic insects.

MUD PUDDLING Temporary puddles are important for insects, and butterflies can be seen "mud puddling" along wet paths. Groups of butterflies gather to feed on carrion, dung, rotting plant matter, and mud, sucking up the fluids through their proboscis (p.16). Male butterflies from the families Papilionidae and Pieridae are particularly attracted to these places, where the salts they obtain help boost their reproductive success and are transferred to the female in the spermatophore (sperm capsule) during mating.

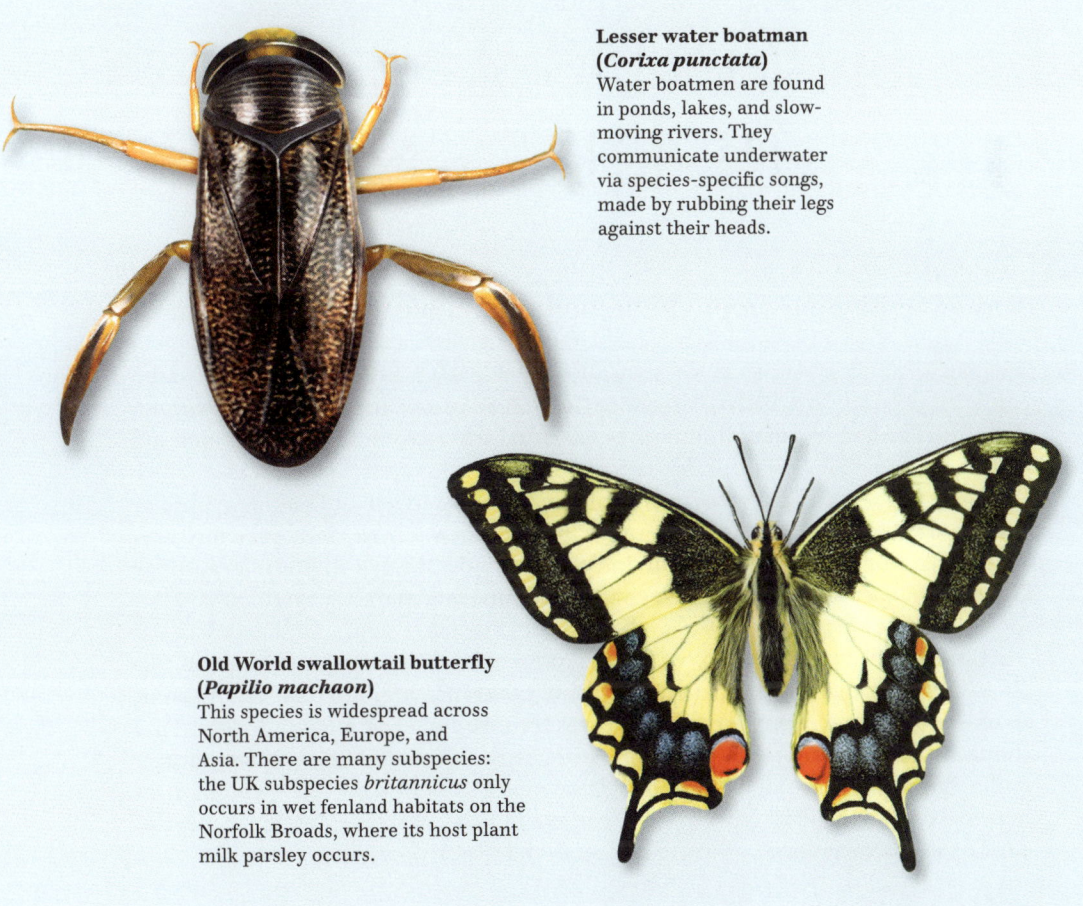

Lesser water boatman (*Corixa punctata*)
Water boatmen are found in ponds, lakes, and slow-moving rivers. They communicate underwater via species-specific songs, made by rubbing their legs against their heads.

Old World swallowtail butterfly (*Papilio machaon*)
This species is widespread across North America, Europe, and Asia. There are many subspecies: the UK subspecies *britannicus* only occurs in wet fenland habitats on the Norfolk Broads, where its host plant milk parsley occurs.

Marshes are covered by water for the majority of the time, so life is harsh for the relatively few insect species that live in these inhospitable habitats.

MARSHLAND

SALTWATER MARSHES Saltwater marshes are important coastal buffer zones. They protect inland freshwater areas from being inundated by saltwater from stormy seas and high tides. Approximately 300 insect species have been described from saltwater marshes worldwide, but relatively little is known of the life histories of most of them. Insects are sensitive to environmental stressors, so those that can survive in marshlands can be important bioindicators of marshland ecosystem health.

DARK SPREADWING DAMSELFLIES Some insect species have developed adaptations to cope with marshland conditions. Dark spreadwing damselflies (*Lestes macrostigma*) live in saltwater and brackish marshes, and they occur along the Mediterranean coast of Southern Europe and into Central Asia and the Middle East. The female lays her eggs on rush and sedge plants, using her ovipositor to cut the plant and deposit her eggs within the protection of the plant tissues. After hatching, the predatory nymphs often feed on mosquito larvae, so they are important for controlling insect vectors of disease. Dark spreadwing damselflies are threatened by habitat modification and the draining of marshlands, as well as by activities to control mosquitoes, which deprive them of one of their favorite prey.

MOSQUITOES *Anopheles* mosquitoes carry malaria, so they can be harmful to people. Campaigns to reduce malaria in the 1960s and 1970s targeted the eradication of this species in its marshland habitat. Not only did these efforts often fail, but they also affected other harmless insects, as marshland habitats were destroyed and pesticides such as DDT caused pollution. In the UK, control activities targeted salt marsh mosquitoes (*Ochlerotatus detritus*), which do not

carry disease but can be problematic, biting people living in urban areas near marshes. Salt marsh mosquitoes lay their eggs in wet sand. However, even if the sand dries out, their eggs can survive for months or years, until conditions improve and the mosquitoes can complete their life cycle.

CONSERVING MARSHLAND In the past, marshlands were often seen as undesirable places, home to mosquitoes that are a nuisance at best and cause malaria at worst. Nowadays, there is more appreciation for the importance of marshland habitats as a home for biodiversity. Marshlands also play a key role in storing carbon—to help alleviate climate change—and controlling flooding. They can act as natural water purifiers, too, filtering out silt and pollutants. Environmentally friendly practices can help conserve marshland, for example, by controlling mosquitoes using natural biological agents, such as *Bacillus thuringiensis* subspecies *israelensis* (Bti). Their toxins specifically target harmful insects instead of killing indiscriminately.

Dark spreadwing damselfly (*Lestes macrostigma*)
The nymphs of this species are associated with vernal ponds, temporary pools of water that dry out in late spring or early summer. Their temporary nature makes them attractive because they tend to lack the fish and other predators more common in permanent waters.

Insects are hard to find in marine environments because most cannot survive in saltwater. Some insects do live in marine habitats, but none are known to spend their entire life cycle fully submerged there.

MARINE HABITATS

Only about 3 percent of insects occur in water, and only a tiny fraction of these are found in marine habitats, such as salt marshes, mangroves, coral reefs, and the ocean. The most common groups are beetles (pp.170–171), flies and mosquitoes (pp.174–175), and true bugs (pp.160–161), which together make up about 75 percent of all marine insects.

INSECTS AND SEAWEED Many rove beetles and tiger beetles are found in seaweed washed up on beaches, coastal salt marshes, or sand dunes. Seaweed flies breed within the piles deposited and can occur in large swarms. Their life cycle is linked to the high spring tides that happen twice a month, associated with full and new moons, and the adults must develop quickly or risk being carried back out to sea at the next spring tide. Tiny midges (such as *Clunio* spp.) are found on shores that reveal seaweed-covered boulders at low tides. They must also complete their life cycle between successive spring tides. The most curious of all marine midges are the flightless *Pontomyia*. Their egg and larval stages are spent entirely underwater among marine plants, and adults emerge mostly at night to coincide with full or new moons, which may help them find mates.

MARINE BUGS The most commonly encountered marine insects are true bugs. Coral treaders (*Hermatobates* spp.), for example, live in air-filled holes in boulders or rock walls along the shores of tropical islands. They come out to feed at low tide, but if they are trapped by the rising tide, they can survive on the sea's surface until the tide turns. Water striders or sea skaters (such as *Halobates* spp.) are found in tropical mangroves. Although they have no wings, they are difficult to catch because they are able to move fast (about 3 ft/1 m per second) and jump high (2 in/5 cm or more). Five species of water striders are the only insects that spend all their lives on the open ocean, hundreds of miles away from land, but how these tiny insects can survive stormy seas and avoid the harmful effects of the sun is not known.

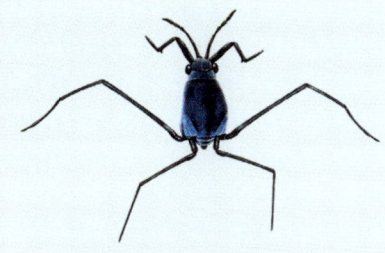

Water strider (*Halobates micans*)
This species has an unusual distribution, occurring in the warmer waters of the Atlantic, Indian, and Pacific Oceans.

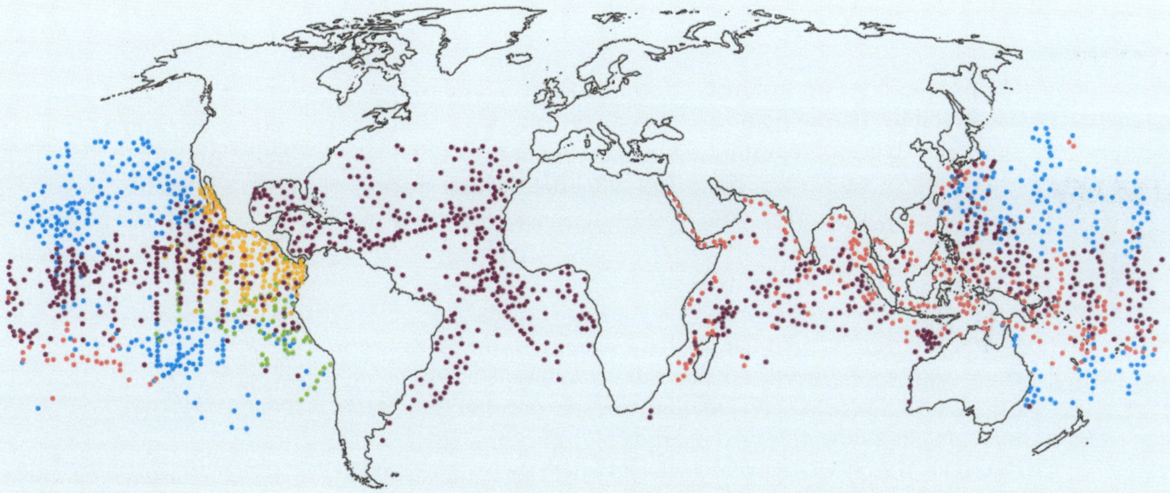

World distribution of water striders
This map shows the estimated global distribution of *Halobates* species, based on more than 3,000 samples from various oceanographic expeditions over the past 50 years.

Key
- ● *Halobates micans*
- ● *Halobates sericeus*
- ● *Halobates germanus*
- ● *Halobates sobrinus*
- ● *Halobates splendens*

Insects living on mountains are adapted to the cooler conditions found on mountaintops. However, high mountain insects—including species of flies, butterflies, and beetles—are threatened by climate change as their habitat warms up.

A. Mountain clouded yellow (*Colias phicomone*) This butterfly has to be tough to survive, and adults are usually very hairy to insulate them against the cold. The short mountain summers mean that it can take two years for them to complete a full life cycle.

MOUNTAINS

HIGH LIFE The start of winter is signaled in temperate regions by the first dusting of snow on the mountains and temperatures starting to drop to below freezing at night. In the high mountains, above 5,900 ft (1,800 m), insects spend the winter as larvae or eggs until spring arrives. For example, in Europe, the eggs and larvae of the Gavarnie blue butterfly (*Agriades pyrenaicus*) survive the winter on the ground or attached to their host plants, protected by a blanket of snow. Underneath the snow, the temperature is about 32°F (0°C), providing relatively warm and stable conditions.

By early summer, the snow starts to melt, triggering the start of the new season. The race begins for high mountain insects to complete their life cycles over the next couple of months. By midsummer, adult butterflies are emerging, and the richness and abundance of mountain butterflies reach their peak. However, temperatures drop again from the beginning of fall, which means another year is over for these insects.

MOUNTAIN SPECIALISTS Butterflies such as the turquoise blue (*Polyommatus dorylas*) and Apollo (*Parnassius apollo*) are mountain specialists. But climate changes mean that winter snow cover is becoming thinner, and it, therefore, offers their eggs and larvae protection for a shorter time. Such environmental shifts are making it much harder for these insects to survive harsh conditions.

A.

B. Gavarnie blue butterfly (*Agriades pyrenaicus*)
This butterfly is found in rocky habitats at about 6,560 ft (2,000 m).

B.

C. Dung beetle (*Trypocopris pyrenaeus*)
In the Picos de Europa mountains, Spain, dung beetles occur in shaded habitats (woodland and heaths) at lower elevations where temperatures are hotter, and in more open habitats (pastures) at colder, higher altitudes.

C.

Some insects have adapted well to city life,
living in park and domestic gardens and on the shoulders
of roads. But urban areas need to be managed better
to help boost insect numbers.

URBAN ENVIRONMENTS

ADAPTING TO CITY LIFE With their huge expanses of concrete, urban areas can be inhospitable places for wildlife, but some insects have adapted quickly to live alongside people in heavily modified landscapes. Pollinators make up a large portion of insects in cities, with some bees preferring urban environments to rural areas. Buildings offer shelter and warmth for insects to undergo diapause (p.136) and are warmer than the surrounding countryside. The red mason bee (*Osmia bicornis*) female makes a nursery for her eggs in holes in house bricks and cracks in mortar and feasts on pollinated garden flowers and fruit trees. Installing bug hotels can help attract solitary bees to gardens.

STRUGGLES IN THE CITY Moths are important nighttime pollinators, but they are distracted by street lighting. This light pollution disrupts their activities, because it attracts the moths toward the light and away from pollination and egg laying. Insect host plants are threatened in towns and cities because they are often seen as untidy weeds and are removed to make room for ornamental plants that insects may not be able to use. For example, bramble

(*Rubus fruticosus*) is crucial to some insects because it flowers late into fall, providing nectar and shelter when other plants are scarce. Caterpillars of the peach blossom moth (*Thyatira batis*) rely on bramble leaves for food from July to September. Moths and butterflies are a regular sight in urban environments where people leave areas to go wild. It is important that we appreciate how crucial these untidy areas are for insects. Rewilding areas and growing native plants give urban insects a helping hand.

**Peach blossom moth
(*Thyatira batis*)**
Larger peach blossom moth caterpillars gorge on brambles during the night and hunker down into leaf litter during the day to evade predators.

Islands that have been isolated from the mainland for millions of years have evolved insects that occur nowhere else on Earth. These species showcase unique adaptations to island life.

ISLANDS

CARNIVOROUS CATERPILLARS Hawaii is a group of volcanic islands that is home to insects with a unique way of feeding. There are 18 species of moths (*Eupithecia*) whose caterpillars ambush their prey. These killer caterpillars rest inconspicuously on the edges of leaves and lie in wait for unsuspecting arthropods (such as cockroaches, crickets, and springtails) to pass by before mounting a deadly attack. The caterpillars use their elongated legs to snatch their prey, then use their strong jaws to tear apart the fleshy food. It provides them with a more nutritious and protein-rich diet than would be obtained from eating plants.

ANTARCTIC ISLAND INSECTS Antarctica has some of the coldest, driest, and windiest islands on Earth, and the continent is home to only one endemic insect species. This flightless midge (*Belgica antarctica*) is found on the western Antarctic Peninsula and South Shetland Islands. It is well suited to extremely cold conditions (pp.136–137), and its larvae can lose up to 70 percent of the water inside their body to prevent ice crystals from forming and damaging their cells. They can produce heat shock proteins and antioxidant enzymes to provide protection from the high levels of ultraviolet radiation produced by the Antarctic sun. They can also create antifreeze compounds (glucose, lipids) that protect them from freezing. This suite of features has taken 40 million years of evolution to perfect, showcasing one of the most astonishing examples of adaptation to life on cold islands.

SWALLOWTAIL BUTTERFLIES Tropical island butterflies, such as those found on Sulawesi, Indonesia, encompass a wide range of colors and shapes. Sulawesi has been isolated for more than 40 million years, resulting in the evolution of many different species, and it is a hotspot of biodiversity, with more than 40 percent of its 488 butterfly species found only on this island. Swallowtail butterflies vary in size, from the small green dragontail (*Lamproptera meges*, forewing length ~¾ in/2 cm) to the large Blume's peacock (*Papilio blumei*, ~2⅓ in/6 cm).

Hawaiian caterpillar (*Eupithecia*)
In Hawaii, caterpillars of the genus *Eupithecia* eat other invertebrates. These carnivorous caterpillars camouflage themselves on plants, waiting for passing prey, then use their long forelegs to snatch and grip the prey before eating it.

Some insects thrive in desert environments. Remarkably, they are able to find food and water, reduce the water loss from their bodies, and avoid or tolerate the heat.

DESERTS

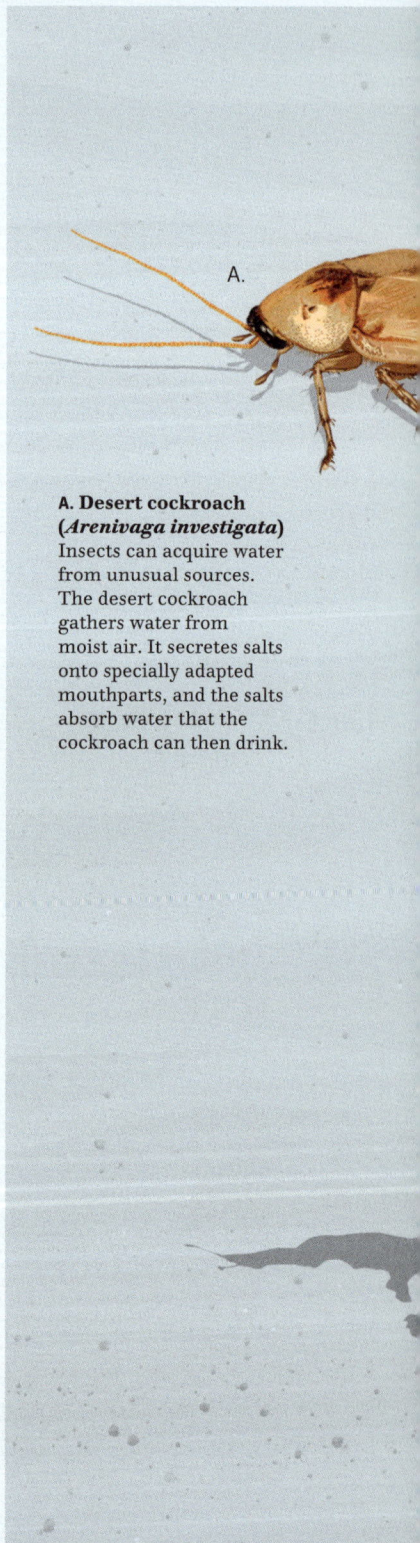

A. Desert cockroach (*Arenivaga investigata*) Insects can acquire water from unusual sources. The desert cockroach gathers water from moist air. It secretes salts onto specially adapted mouthparts, and the salts absorb water that the cockroach can then drink.

The first defense for desert insects against harsh conditions is their behavior. Many spend the day in burrows, where humidity is high, thus avoiding surface temperatures that can exceed 122°F (50°C).

COPING WITH THE HEAT Sahara desert ants (*Cataglyphis bicolor*) can tolerate very high body temperatures (above 122°F/50°C) that would kill other insects. They use precise navigation to minimize the distance they travel, foraging for insects that have been killed by heat on the dunes. In the deserts of Southwestern North America, the tobacco hornworm (*Manduca sexta*) lays its eggs on the underside of leaves of deep-rooted plants. These plants have access to underground water, and evaporation from the leaf cools the air and surrounds the eggs in a humid pocket of air. Dung-ball-rolling beetles in Southern Africa prevent themselves from heating up too much when they are on very hot ground by standing on their dung ball from time to time to cool their feet.

COPING WITH DRY ENVIRONMENTS Here, insects have several ways of surviving. Some use their stored carbohydrates like a sponge to store water. Having a waterproof cuticle helps reduce water loss, and many desert darkling beetles secrete a waxy waterproofing layer (bloom). Other insects are very efficient at absorbing water during digestion, producing very dry feces and very little urine.

B. Fog-basking beetle (*Onymacris unguicularis*)
This beetle has a specially adapted cuticle that collects droplets of water from the fog that blankets the desert at night. It then drinks this water by lowering its head so the water rolls down its elytra (hardened forewings) and into its mouth.

B.

C.

C. Dung-ball-rolling beetle (*Scarabaeus lamarcki*)
This dung beetle is standing on its dung ball to escape the hot ground. Researchers experimentally fitted dung beetles with tiny protective boots and found that the insects spent less time on their dung balls protecting their feet from the heat.

*Insects are cold-blooded (ectotherms) and cannot regulate
their body temperature. Nonetheless, they are able to
thrive in some of the coldest places on Earth.*

VERY COLD ENVIRONMENTS

WINTER DORMANCY Insects in cold places often spend the coldest part of the year in a dormant state, known as diapause. They usually become dormant at the stage of their life cycle, such as egg or larva, that is particularly good at tolerating the stresses of winter. When dormant, their metabolic rates are lower, which helps save energy for development when conditions warm up in spring.

ACTIVE IN THE COLD A few insects continue to thrive when snow is on the ground. For example, ice crawlers in the mountains of western North America and Asia are active on snow and ice, scavenging for other insects that get blown onto the snow from lower altitudes. Other species move around in the space between the snow and the ground, where temperatures are more constant and relatively mild. Sometimes, they come to the surface, most notably the species of dark springtails known as snow fleas. (Springtails are no longer classified as insects.) These can be found swarming in northern forests, even in midwinter.

ANTARCTIC MIDGES The only insect in Antarctica is the Antarctic midge (*Belgica antarctica*), which spends most of its life as a larva. It can tolerate freezing conditions, and adults are present only briefly in the summer. About a dozen species of springtails also occur in Antarctica, where they live on small patches of ice-free land at the coast. Springtails can survive extremely low winter temperatures—by avoiding being frozen—and some remain active at temperatures below 23°F (-5°C). Antarctic summers have 24-hour sunlight, which heats up dark rocks. Consequently, some Antarctic springtails also have to be able to tolerate surprisingly high temperatures—above 95°F (35°C) in some cases.

SURVIVING EXTREME COLD Insects that encounter very low temperatures survive by either preventing ice from forming in their bodies (freeze avoidance) or withstanding being frozen (freeze tolerance). Freeze-avoiding insects accumulate chemicals in their hemolymph (equivalent to blood), which prevent ice crystals from forming. The emerald ash borer beetle (*Agrilus planipennis*) overwinters as a prepupa and usually freezes at around -22°F (-30°C). However, after experiencing a polar vortex in the Canadian prairies, it could stay unfrozen as low as -58°F (-50°C). Some freeze-tolerant insects are extraordinarily cold-hardy. Malt fly (*Chymomyza costata*) larvae can survive immersion in liquid helium, which at -440°F (-262°C) is colder than almost anywhere in the universe.

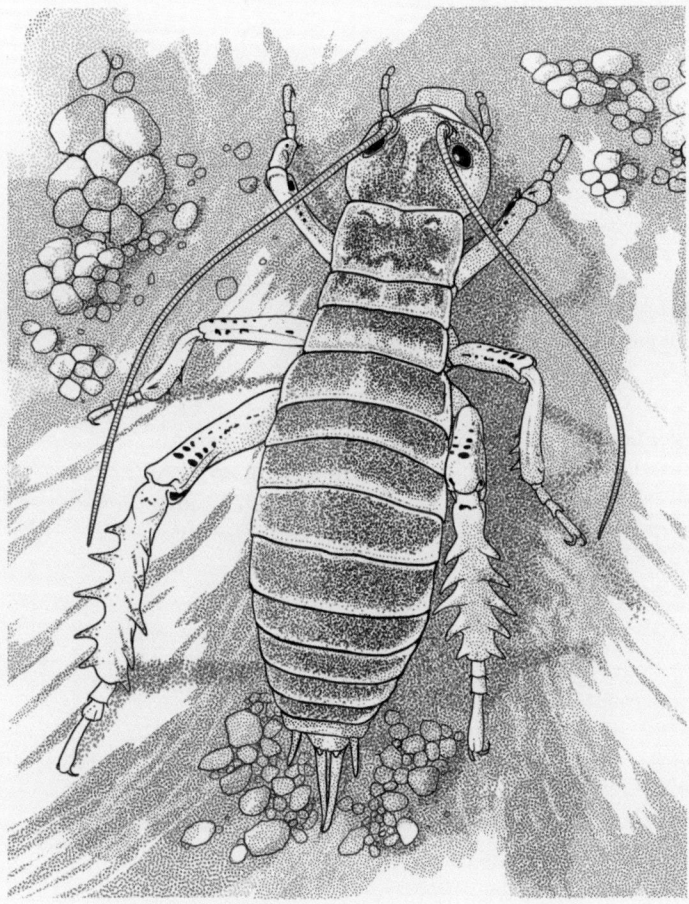

New Zealand alpine wētā (*Hemideina maori*)
The New Zealand alpine wētā is freeze tolerant and can survive when 80 percent of the water in its body is frozen.

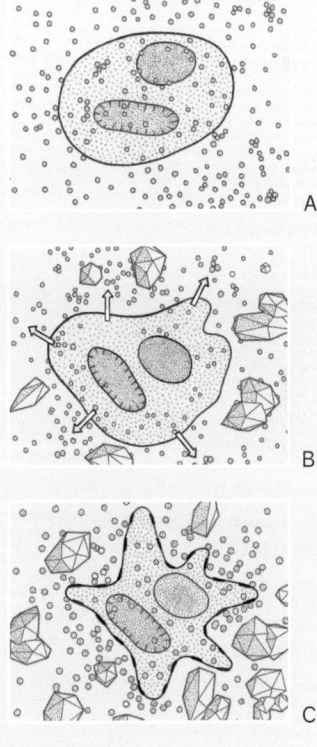

Extracellular ice formation
Freeze-tolerant insects appear to prevent ice from forming inside their cells.
A. Concentrations are usually the same inside and outside the cell.
B. Water freezes in the hemolymph. The increased concentration draws water from the cell.
C. When the insect is fully frozen, the dehydrated cell is so concentrated that it remains unfrozen.

There is a hidden world of caves in limestone areas around the world that harbor a rich variety of weird and wonderful insects, known as troglobites, which live away from the light.

CAVES

Animals that are found in caves live in constant darkness, and their food is often scarce. Nonetheless, several arthropod groups have evolved to survive in this environment, such as insects, spiders, millipedes, and centipedes. Even though these animals are not closely related, they share similar characteristics (so-called "convergent evolution"). Many of them are pale, because they lack pigment; they do not have eyes; and they develop elongated limbs.

ENTRANCE, TWILIGHT, AND DARK ZONES Caves are divided into three zones, with different types of insects associated with each one. Sunlight can penetrate the entrance zone, which is mainly inhabited by temporary visitors from the outside world who cannot live within cave habitats. The twilight zone does not experience direct sunlight, but it is influenced by the outside world, with fluctuations in humidity and temperature. This zone is occupied by species that only use caves for part of their lives, such as overwintering moths and gnats. It is also visited by insects that move in and out of caves on a daily basis, such as many cave crickets. Troglobites live in the dark zone, where they experience constant darkness, as well as constant temperatures and very high humidity. Their food is often extremely limited. However, in tropical regions, insects often share caves with large populations of bats, whose guano provides food for many detritivores (detritus eaters), such as cave cockroaches. Undoubtedly, there are still numerous troglobites awaiting discovery, particularly in tropical caves that are currently understudied.

A. Glow worm (*Arachnocampa*) These larvae dwell in the twilight zone. Their bioluminescence lights up the caves and attracts prey, such as moths and other flying insects, which they catch in sticky silk snares.

ENTRANCE ZONE

B. Hawaiian cave planthopper (*Oliarus polyphemus*)
This cave insect is found only in Hawaii, where it lives in the twilight zone in crevices. Nymphs feed by sucking the sap from tree roots that penetrate the rock face.

A.

B.

C.

DARK ZONE

TWILIGHT ZONE

C. Cave beetle (*Leptodirus hochenwartii*)
The oddly shaped detritivore cave beetle—with its very long, slender thorax and domed abdomen—is found in the dark zone of caves in Slovenia and Croatia.

Six-spot burnet (Zygaena filipendulae)

CLASSIFY

THE ESTIMATED 5.5 MILLION insect species on Earth have been divided into 27 major taxonomic groups. Discover what separates a dragonfly from a dobsonfly or a snakefly from a scorpionfly, and explore examples of insects from each of the 27 groups.

Insects are classified into approximately 27 orders,
but the diversity of insects is so great that it is not known
how many species there are.

THE DIVERSITY OF INSECT ORDERS

The amazingly diverse form and natural history characteristics of each order of insects have been maintained for millions of years, despite the evolution of many modifications as species adapted to new environments. These modifications have allowed insects to live in virtually every terrestrial and freshwater habitat on Earth, as well as in some marine habitats. They are absent only from the deep sea.

The most species-rich orders, with more than 100,000 described species each, are the beetles (Coleoptera); flies (Diptera); butterflies and moths (Lepidoptera); and ants, bees, and wasps (Hymenoptera). However, some of the other minor orders, such as earwigs (Dermaptera) and fleas (Siphonaptera), are highly diverse compared to other animal groups. There are about 2,500 species of fleas, for example, which is about the same number as all the world's rodent species combined.

DIVERSITY WITHIN THE ORDERS Within each insect order, entomologists group species in families. The true bugs (Hemiptera), for example, contains more than 130 insect families, including aphids or greenflies (5,500 species) and cicadas (3,000 species). In contrast, there are far fewer species—about 50—of adelgids. This family is closely related to the aphids but lacks the tail-like structure that aphids have. The beetles order is currently the largest insect order by a sizable margin, and more than 95,000 species are known from just one family: the weevils. This insect family, therefore, contains more species than all the birds, mammals, and other vertebrates combined. Although the

beetles are the most diverse in terms of numbers of described species, the huge number of undescribed species among the small flies and parasitoid wasps has led many scientists to suspect that the flies and the ants, bees, and wasps might be the largest orders if all existing species were discovered.

THE FINAL FRONTIER New species of insect are discovered far more frequently than in most other animal groups. Recent studies using DNA have enhanced the ability to separate and characterize many species that in previous years appeared indistinguishable.

Hemiptera: large milkweed bug (*Oncopeltus fasciatus*)

NUMBER OF DESCRIBED SPECIES IN THE WORLD

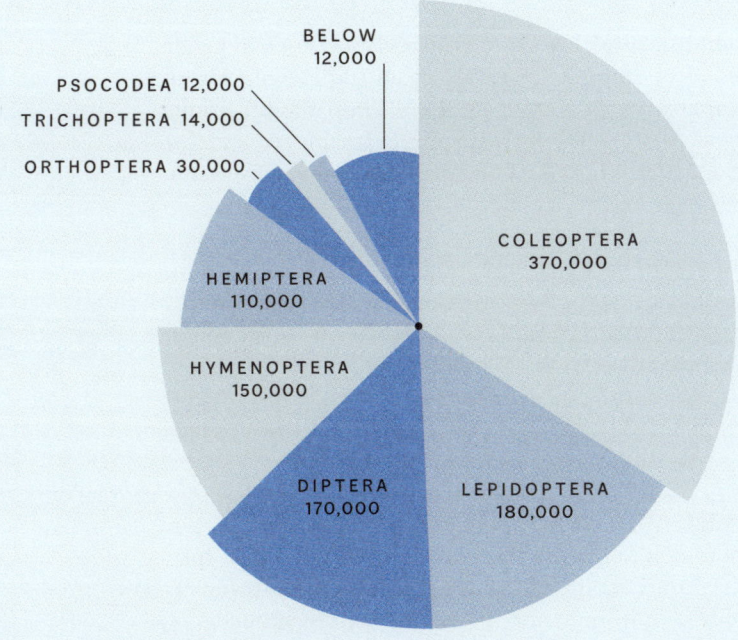

BELOW 12,000

PSOCODEA 12,000

TRICHOPTERA 14,000

ORTHOPTERA 30,000

HEMIPTERA 110,000

HYMENOPTERA 150,000

DIPTERA 170,000

LEPIDOPTERA 180,000

COLEOPTERA 370,000

Orders with less than 12,000 described species
Thysanoptera, Neuroptera, Plecoptera, Ephemeroptera, Phasmida, Zoraptera, Blattodea, Mantodea, Siphonaptera, Dermaptera, Mecoptera, Strepsiptera, Odonata, Zygentoma, Archaeognatha, Embiidina, Megaloptera, Raphidioptera, Notoptera

Hymenoptera: European beewolf (*Philanthus triangulum*)

Lepidoptera: six-spot burnet (*Zygaena filipendulae*)

*Bristletails appeared in the fossil record in the Devonian period
(about 417 to 354 million years ago) around the same time
as spiders and scorpions. Worldwide, there are around
500 known species in two families.*

ARCHAEOGNATHA
BRISTLETAILS

MACHILOIDES
Rock bristletail

FAMILY Meinertellidae

SPECIES IN FAMILY Around 170

HABITAT Rocks, leaf litter, tree trunks, tree canopy

Bristletails have long terminal filaments at the end of their abdomen, which, together with a pair of cerci (p.148), give them a three-tailed appearance.

ANCIENT INSECTS Bristletails belong to a group of wingless insects known as Apterygota, and they get their common name from the three flexible, bristlelike filaments at the tip of the abdomen. They do not undergo metamorphosis (pp.26–27), but instead molt continuously, even after reaching sexual maturity. Together with silverfish and firebrats, bristletails are some of the most primitive insects, and all three have physical similarities, such as a cylindrical body shape and no wings.

HIGH JUMPERS Bristletails have a small head and long body, which is arched upward at the thorax. If disturbed, some species can spring up to 12 in (30 cm) into the air by suddenly flexing their abdomen. Bristletails' eyes are very conspicuous because they are located on top of the head, touching each other, and their mouthparts are partly retractable, with simple chewing mandibles. The seven segments of the mouthparts (maxillary palps) are very long, often longer than the insect's legs. Bristletails have a thin exoskeleton that offers little protection against dehydration, so they are found in moist places—such as under stones or bark— where they feed on algae, lichens, mosses, or decaying organic matter. They mate once at each adult instar and live for up to four years, which is longer than most insects.

*There are less than 550 known species
of silverfish and firebrats in five families. Silverfish get
their common name from the easily detachable scales that
cover their body and can appear shiny.*

ZYGENTOMA
SILVERFISH AND FIREBRATS

PRIMITIVE INSECTS Like bristletails, silverfish and firebrats are wingless and do not undergo metamorphosis (pp.26–27). They have also been around for millions of years, coping with the considerable global environmental climate and habitat changes that have occurred over time. They are superficially similar in appearance to bristletails by having three filaments on the end of their abdomen, but they are different because the filaments often stick out sideways from the body, sometimes even at right angles. The bodies of silverfish and firebrats are also less cylindrical, tapering toward the rear to be more spindle-shaped, and their characteristic three-tailed appearance is made up of a long central filament and two shorter ones on each side. Most species have eight pairs of short appendages (styli) on their abdomen, although some have fewer appendages or none at all.

FAST RUNNERS Unlike those of bristletails, the eyes of silverfish and firebrats are small or absent, and there are no ocelli (simple eyes). They also have much shorter mandibles, and the five segments of their mouthparts (maxillary palps) are of normal length. Silverfish can run fast, but they lack the jumping ability of bristletails. However, their slippery scales probably give them similar protection from predators as they make them difficult to hold onto. Silverfish feed on organic detritus and plant matter with a high cellulose content, and some species have become pests in people's houses because they eat paper, cloth, and books and also contaminate food.

LEPISMA SACCHARINUM
Common silverfish

FAMILY Lepismatidae

SPECIES IN FAMILY Around 190

HABITAT Moist places, including in homes and other buildings

The presence of cellulase, produced by the midgut, allows silverfish to digest cellulose.

There are about 7,000 species in this order. Fossils first appeared some 250–200 million years ago, but the ancestors of today's species date back to about 300 million years ago, making them one of the earliest groups of winged insects.

ODONATA
DRAGONFLIES AND DAMSELFLIES

CORDULEGASTER BOLTONII
Golden-ringed dragonfly

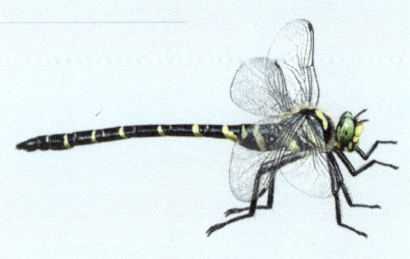

FAMILY Cordulegastridae

SPECIES IN FAMILY Around 55

HABITAT Spring brooks, mainly in forests

Dragonfly nymphs live for three to five years buried in the stones and leaf litter of shallow streams, barely covered by water.

ACROBATIC IN THE AIR Dragonflies and damselflies are the most conspicuous and magnificent insects to be found in freshwater ecosystems. Adults are often brightly colored, with iridescent or metallic hues, and on warm summer days they can be seen flying at amazing speeds to catch insects or flitting around the water's edge looking for prey and suitable mates. Damselflies and dragonflies can be differentiated by the shape of their wings. The hind wings of dragonflies are much broader than their forewings, whereas both pairs of wings of damselflies are similar. Damselflies are usually smaller and more delicate, but in the tropics, there are very large damselflies and tiny dragonflies.

SUPERB PREDATORS Both nymphs and adults are predators. Nymphs live underwater perfectly hidden, buried in the riverbed, clasping to rocks, or sprawling between aquatic plants. There, they wait for unsuspecting prey, which they catch by projecting their prehensile mouthparts at lightning-fast speed. The lower jaw (labium) has hooks and spines, which also help catch prey. Nymph development takes a few weeks to several years, and the adults emerge and leave the water. The winged adults are superb fliers, catching prey in flight by using their legs as a "basket." They have excellent eyesight—their large eyes take up almost the entire head—and their strong jaw makes them dangerous aerial predators.

Worldwide, there are around 4,000 species of stoneflies in 17 families. Early stonefly fossils date back to about 300 million years ago.

PLECOPTERA
STONEFLIES

INDICATORS OF WATER POLLUTION Stonefly eggs and nymphs develop in freshwater environments, and both nymphs and adults can be recognized by a pair of long cerci protruding from the tip of their abdomen. All species are sensitive to water pollution, so their occurrence can be used as an indicator of good water quality. The nymphs are usually found only in fast-running water because they require high oxygen concentrations. Initially, they eat living or dead plant material, but older nymphs also prey on small invertebrates.

The duration of the nymphal stage varies between species and is affected by temperature, with emergence being more rapid in warmer conditions. Most species have a life cycle of a single generation per year, although some are known to live for up to three years as nymphs before emerging as an adult.

DRUMMING TO ATTRACT MATES Adults feed very little, if at all, although some species feed on lichens and algae. Many adult stoneflies are active during spring. They are short-lived and reluctant to fly, so they are often found running around among stones and vegetation on riverbanks. When adults do fly, the movement is often slow and pretty erratic, and some species are wingless or only have very small wings (brachypterous). Stoneflies are generally nocturnal, and males and females communicate by drumming their abdomen on the ground using a pattern that is specific to each species. This drumming results in mating.

EUSTHENIA NOTHOFAGI
Otway stonefly

FAMILY Eustheniidae

SPECIES IN FAMILY Around 30

HABITAT By streams in temperate rainforest

This species is only found in the Otway Ranges of Australia, where the myrtle beech tree (*Nothofagus cunninghamii*) occurs (hence the insect's scientific name). It was previously thought to have gone extinct.

*Grasshoppers, crickets, and bush-crickets evolved more than
300 million years ago, and there are currently close to 30,000 described
species in about 20 families. They often occur in grassland, where the
sounds some species make are evocative of summer.*

ORTHOPTERA
GRASSHOPPERS, CRICKETS, AND BUSH-CRICKETS

Grasshoppers, crickets, and bush-crickets can be easily recognized by their saddle-shaped pronotum, which is a platelike structure that forms part of the thorax, directly behind the head, and often covers the base of the wings. As well as crickets and grasshoppers, the order includes insects that are mimics and predators (for example, many tropical bush-crickets), and also those that are silk spinners (raspy crickets) and blind cave dwellers (cave wētās). Some species (for example, the desert locust *Schistocerca gregaria*) are pests and cause considerable damage to crops, such as maize.

SINGING Many Orthopteran insects make sounds by stridulation, which involves the insect rubbing one part of its body against another. Using wings equipped with small pegs along one of the wing veins and a hardened edge for scraping, crickets and bush-crickets produce the most diverse range of sounds of the order, including some of the highest-pitched sounds in nature. Their raspy chirps and repetitive trills are mostly produced by males to attract females, but they are also used for scaring predators and fighting rivals.

LIVING THERMOMETERS The number of times a cricket chirps increases with hotter weather. Consequently, songs of crickets have often been used to calculate temperature, a rule known as Dolbear's law. Counting the number of chirps a field cricket (*Gryllus bimaculatus*) produces in 15 seconds and adding 40 supposedly gives the temperature in Fahrenheit.

SUPERSONUS AEQUOREUS

Supersonic bush-cricket

FAMILY Tettigoniidae

SPECIES IN FAMILY Around 8,000

HABITAT Widespread

This species makes the highest-pitched sound of any known Orthopteran. It is higher pitched than bat echolocation and can only be recorded with special ultrasonic microphones.

GRYLLOTALPA GRYLLOTALPA

European mole cricket

FAMILY Gryllotalpidae

SPECIES IN FAMILY 137

HABITAT Underground

Mole crickets live in burrows, where the females care for their young. Males construct these burrows with a horn-shaped opening, which functions like a megaphone to amplify their song.

SCHISTOCERCA GREGARIA

Desert locust

FAMILY Acrididae

SPECIES IN FAMILY Around 7,000

HABITAT Various, including grasslands and crops

The desert locust is arguably the world's most destructive migratory pest, its swarms capable of moving hundreds of miles in a day.

GRYLLUS BIMACULATUS

Field cricket

FAMILY Gryllidae

SPECIES IN FAMILY Around 3,500

HABITAT Various, including grasslands and forests

Male field crickets are aggressive to each other and may fight over living spaces or females by gripping their opponent's mandibles.

HEMIDEINA THORACICA

Auckland tree wētā or tokoriro

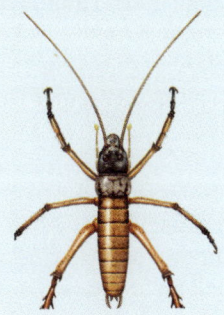

FAMILY Anostostomatidae

SPECIES IN FAMILY Around 300

HABITAT Various, including forests and grasslands

Tree wētās are only found in New Zealand, feeding during the night and resting in tree cavities during the day.

TACHYCINES ASYNAMORUS

Greenhouse camel cricket

FAMILY Rhaphidophoridae

SPECIES IN FAMILY Around 900

HABITAT Mainly forests and caves

This camel cricket is native to Asia but is now commonly found in heated greenhouses in other parts of the world.

In a group as hyperdiverse as insects, angel insects
stand out as one of the least diverse, with only 51 species
divided into two families, four subfamilies,
and nine genera.

ZORAPTERA
ANGEL INSECTS

ZOROTYPUS WEIWEII
Angel insect

FAMILY Zorotypidae

SPECIES IN FAMILY Around 50

HABITAT Tropical rainforest

This species has only recently been found in the rainforest of Borneo, and it was named after the Chinese entomologist Zhang Weiwei.

LIVING ARRANGEMENTS All species of angel insects are small, often ⅛ in (3 mm) or less in length. What they lack in size, they make up for in their fascinating life history. Angel insects live gregariously in small colonies (15–120 individuals) in crevices under the bark of rotting logs and dead wood. In this habitat, they frequently groom each other to remove fungal spores or hyphae (long filaments) that might otherwise infect the colony. They feed on these fungal spores and hyphae, but may also scavenge for nematodes (roundworms or eelworms), mites, or other minute arthropods.

NEW COLONIES Individuals are generally pale in color, superficially resembling termites (which they are not related to). They are whitish as nymphs but become a more reddish brown as they mature. While living in the colony, angel insects are blind and wingless. However, when the colony becomes crowded or the log is no longer suitable, new individuals are born with fully developed eyes and wings. These winged insects disperse to mate and produce new colonies.

*There are fewer than 60 known species of gladiators and ice crawlers.
They are united within the superorder Notoptera, which comprises
five families, three of which are known only from fossils
and have no living representatives.*

NOTOPTERA
GLADIATORS AND ICE CRAWLERS

NEW ORDER Gladiators are the newest order of insects, and they were only described in 2002. Ice crawlers are closely related to gladiators, but they were described longer ago in 1914. The two share common characteristics, such as a soft body and no wings.

INTERESTING FEET Gladiators are known as heelwalkers due to their unusual feet. These have a large, padded lobe (arolia), which the insects hold upright, thus giving the appearance of walking on their heels. Found in Southern and Eastern Africa, gladiators are predators and have raptorial legs like praying mantises (p.158) to catch and hold their prey.

LOVE THE COLD Ice crawlers are small insects (⅗–1¼ in/ 15–30 mm) with reduced or absent compound eyes and long antennae. Usually found in leaf litter, they feed on a wide variety of things, including dead animal and plant matter and other insects, and are active at night. They are especially interesting because, unusually for insects, they live in cold, high-altitude environments throughout Asia and North America. Because of where they occur, many species will be under threat from the effects of a warming planet, and they cannot withstand temperatures outside the range of 34–39°F (1–4°C).

*MANTOPHASMA
ZEPHYRA*
Heelwalker or gladiator

FAMILY Mantophasmatidae

SPECIES IN FAMILY Around 20

HABITAT Dry deserts and grasslands

This species is found only in Namibia. Its species name *zephyra* is Latin for "west wind."

Occurring on every continent except Antarctica, stick insects currently boast a diverse array of some 3,500 species. They are masters of disguise in the insect world, employing camouflage to seamlessly hide among the plants they inhabit.

PHASMIDA
STICK INSECTS

Stick insects, also called stick-bugs, walkingsticks, or stick animals, are a group of plant-eating insects that are easily recognized by their elongated and twiglike appearance. They are classified within the order Phasmida, named from the ancient Greek word *phasma*, meaning "phantom" or "apparition," which reflects their impressive ability to look like a plant instead of an insect.

PLANT MIMICS Many stick and leaf insects are remarkably elusive, mastering the art of camouflage in their natural habitat. This is tailored to the plant that each insect is associated with, and stick insects have evolved to resemble pine needles, branches, or leaves. Some species have improved their camouflage by developing spines and outgrowths to break up their outline or by evolving color blotches or texturing to mimic the damage that leaves suffer when they are nibbled. Stick insects also conceal themselves from predators by swaying gently when disturbed to mimic the way that plants move in the wind. This motion camouflage adds another layer of deception, making it even more challenging for their enemies to spot them among the foliage.

REPRODUCING The majority of stick insects reproduce via sex between a female and male insect, but some reproduce asexually (parthenogenesis). This means that females lay unfertilized eggs, which subsequently develop and hatch into female offspring. Because a single female can lay hundreds of eggs, this mode of reproduction enables a large population of stick insects to originate from a single female, producing numerous copies of herself. This extraordinary strategy allows stick insect numbers to increase very rapidly, even when males are absent.

PHOBAETICUS CHANI
Chan's megastick

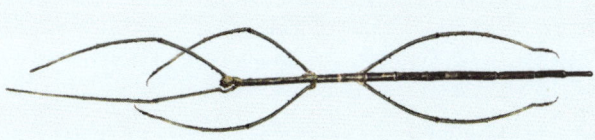

FAMILY Phasmatidae

SPECIES IN FAMILY Around 700

HABITAT Rainforest

At an astonishing 14 in (35 cm) long, female Chan's megasticks are the longest insects in the world yet described.

TIMEMA CRISTINAE
Cristina's timema

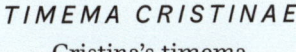

FAMILY Timematidae

SPECIES IN FAMILY 21

HABITAT Chaparral (shrubland)

At just ¾ in (2 cm) long, male Cristina's timema are the shortest stick insects in the world yet described.

PERUPHASMA SCHULTEI
Black beauty stick insect

FAMILY Pseudophasmatidae

SPECIES IN FAMILY Around 320

HABITAT Montane forest, neotropical

This species can defend itself by spraying predators with a defensive liquid from glands at the rear of its head.

CARAUSIUS MOROSUS
Indian stick insect

FAMILY Lonchodidae

SPECIES IN FAMILY Around 1,200

HABITAT Mountainous rainforest

This is a common species to keep as a pet. Females reproduce asexually by parthenogenesis.

The insects in this small order (less than 500 species) are notable for their long body, short legs, and gregarious subsocial behavior.

EMBIOPTERA
WEBSPINNERS

OLIGOTOMA NIGRA
Black webspinner

FAMILY Oligotomidae

SPECIES IN FAMILY Around 60

HABITAT Widespread, ground-dwelling

This particular species is invasive. It is originally from India but is now found across the world.

SUBSOCIAL SPINNERS Webspinners have unique enlarged protarsomeres (the first joint of their feet), which contain a large number of silk glands. They use these to "paint" silk with rapid brushstrokes, creating a long, complex tunnel in which they reside (pp.58–59). Females spend most of their life in these tunnels, caring for their young and maintaining the tunnel network. They rarely disperse, so sizable groups of semirelated webspinners eventually find themselves living together in a subsocial colony. Unlike in fully social insect colonies (pp.46–47), adults provide only limited parental care to their offspring. The winged males disperse to find new colonies in which to settle, mate, and die. The females of some species eat the male soon after mating—a far cry from their usual diet of leaf litter, moss, and lichen.

MORE TO LEARN Distributed worldwide, webspinner species are richest and most abundant in the tropics. However, as these insects have never received significant attention from scientists, there is still much to learn about their systematics (pp.184–185) and behavioral ecology, as well as the effect of subsocial behavior on their evolution.

*Cockroaches and termites are thought to have
evolved alongside the dinosaurs during the Triassic period,
about 200 to 250 million years ago.*

BLATTODEA
COCKROACHES AND TERMITES

Cockroaches and termites together form the Blattodea, an order sister to the predatory mantises. Estimates suggest about 5,000 cockroach and 3,000 termite species in existence.

COSMOPOLITAN CRAWLERS Cockroaches can be found all over the world, in natural habitats and man-made places. They even make their way into household electrical items, such as refrigerators and microwave ovens. Cockroaches come in many sizes and colors. The smallest live symbiotically with ants or termites and measure 1/10 or 1/8 in (2 or 3 mm). They contrast with the gigantic South American *Megaloblatta* cockroaches, which are up to 4 in (10 cm) in length and have a wingspan of nearly 8 in (20 cm). Some species are wingless, such as the Madagascar hissing cockroach (*Gromphadorhina portentosa*), which has evolved a remarkable defense system. When threatened, it "hisses" by expelling air through specialized abdominal spiracles (respiratory openings).

SOIL ENGINEERS Although cockroaches are known to spread bacteria, they also perform important ecosystem services. They eat all types of organic matter (wood, leaf litter, and so on), which they then decompose, thus contributing to maintaining soil health. This role has been perfected by a special kind of cockroach: termites (pp.52–53).

*PERIPLANETA
AMERICANA*
American cockroach

FAMILY Blattidae

SPECIES IN FAMILY 760

HABITAT Various, including leaf litter, rotting wood, and domestic houses

The American cockroach is the largest of the common cockroach species. Despite its name, it is originally from Africa and the Middle East.

Praying mantids are found all around the world,
with more than 2,500 species known to exist. They are predatory
insects with a lightning-quick strike.

MANTODEA
PRAYING MANTIDS

HYMENOPUS CORONATUS
Malaysian orchid mantis

FAMILY Hymenopodidae

SPECIES IN FAMILY Around 300

HABITAT Various, from grasslands to forests

This species mimics an orchid, hence its common name. Family members have various adaptations to deter predators, including forewings decorated to look like an eye.

SKILLED HUNTERS Praying mantids (or "mantises") are probably the best known of the insect predators. The key to their hunting success is their keen eyesight and super-fast strike, in which they quickly grab and subdue their prey with their large, spiky front legs. These highly specialized limbs lend the common name to the group, not because they are "preying," but instead "praying." A close look will reveal how the insects hold their front legs together, as if in prayer.

ON THE MENU Most mantids live in the tropics, where there are plenty of other insects for them to catch and eat. Larger species sometimes consume other animals, including lizards, small birds, and even fish. Mantids are also cannibalistic and will gladly eat smaller individuals of their own kind. Maybe most surprisingly, females sometimes feast on their male suitors soon after mating—unless the males are able to make a quick escape by jumping or flying away.

DEADLY DISGUISE Mantids are masters of disguise, which helps them both avoid predators and surprise their unsuspecting prey. Different species variously resemble leaves, twigs, and even orchid flowers (left)—beautiful but deadly.

*Mayflies are an ancient lineage of winged insects,
with fossil records extending back some 300 million years.
Worldwide, there are around 4,300 known species
in about 40 families.*

EPHEMEROPTERA
MAYFLIES

AQUATIC INSECTS Mayflies are found in freshwater rivers and streams. Nymphs are always aquatic, and most of the life cycle is spent in this stage, with molting occurring up to 50 times in some species. Adult mayflies do not feed, and they live for only a very short time: a few hours to a few days depending on the species. They have a long tail and long wings that do not fold flat over their body, which is characteristic of more ancient insects.

LIFE CYCLE Aquatic mayfly nymphs are easily distinguished from most other aquatic insects because they have three long filaments at the tip of the abdomen and feathery or platelike gills along the sides of the abdomen. Most nymphs graze or gather decomposing fine particulate matter composed of plants and algae, but a few are predators. Nymphs develop into a fully winged but sexually immature stage that resembles the adult. At this stage, they have opaque wings that are covered with small, water-resistant hairs—and relatively short, unfurled forelegs in males. In contrast, adults have clear wings, and males usually have extended forelegs for grasping females during mating. The length of the life cycle varies among species, with some producing several generations each year, but others taking up to two years to complete a generation.

*ECDYONURUS
TORRENTIS*

Large brook dun

FAMILY Heptageniidae

SPECIES IN FAMILY Around 660

HABITAT Shallow, fast-moving rivers and streams

Mayflies are sensitive indicators of environmental change, pollution, and water quality in freshwater habitats.

Worldwide, there are around 2,000 species of earwigs in 11 families.
Dermaptera evolved during the middle to late Jurassic and
diversified over millennia during the Mesozoic Era
between 253 and 66 million years ago.

DERMAPTERA
EARWIGS

FORFICULA AURICULARIA
Common or European earwig

FAMILY Forficulidae

SPECIES IN FAMILY Around 500

HABITAT Dark, damp places under stones and logs

Earwigs use their cerci to catch prey, in fighting and courtship, and when folding their wings under their wing cases after flight.

NATURE'S RECYCLERS Earwigs occur commonly across the world in woodland, fields, and gardens. As omnivores, they eat a wide variety of living and dead animal and plant material, so they play an important role in recycling nutrients within ecosystems. Earwigs are nocturnal and live in dark, humid places in leaf litter; under bark, stones, or logs; and in nooks and crevices. One of their favorite habitats is a compost heap, which provides warmth and shelter as well as somewhere they can hunt for a nutritious and plentiful meal. Such places keep earwigs cool in summer and warm and protected from the cold in winter.

CARING FOR YOUNG Earwigs develop through incomplete metamorphosis (pp.30–31), having a few generations of nymphs that eventually develop into adults. The female earwig is one of the few nonsocial insects that engage in the maternal care of their young, from eggs to nymphs. The mother keeps her eggs warm, clean, and free of fungus, then watches over her young to prevent them from being eaten. Adult earwigs are long (up to ⅗ in/15 mm) and thin in appearance, and they are easily recognized by their characteristic pair of cerci (sensory appendages that look like pincers) on the end of their flexible abdomen. The cerci are curved in males, whereas in females they are straight, and both flex their cerci when threatened to deter predators. Many earwigs rarely fly, and they keep their membranous wings folded away underneath a pair of protective wing cases.

Largely overlooked and understudied despite their large numbers, barklice, booklice, and parasitic lice make up the order Psocodea, which comprises around 12,000 species.

PSOCODEA
LICE

FREE-LIVING LICE Barklice and booklice are similar in size and shape, both being tiny (adults are around 1mm long), soft-bodied insects. They are found within various habitats, including the bark of trees, leaf litter, and stored food products. They serve vital roles in the functioning of ecosystems, contributing to decomposition and nutrient cycling. While booklice and barklice are free living, parasitic lice are flat, wingless insects entirely dependent on their host for food, protection, and movement.

PARASITIC LICE There are two types of parasitic lice: chewing lice, which primarily feed on skin debris, secretions, and feathers; and sucking lice, which have evolved specialized mouthparts designed for piercing and feeding exclusively on blood. Excellent examples of adaptations seen in lice can be found in the three (sub)species of human lice, which have each adapted to thrive in a specific region of the human body: the head louse (*Pediculus humanus capitis*), body louse (*Pediculus humanus humanus*), and pubic louse (*Pthirus pubis*, right). Body lice, for example, have adapted to live and lay their eggs in clothing. They consume more food than head lice, lay more eggs, and grow faster. The smaller head lice, in contrast, are adapted to clinging tenaciously to human hair.

PTHIRUS PUBIS
Pubic louse

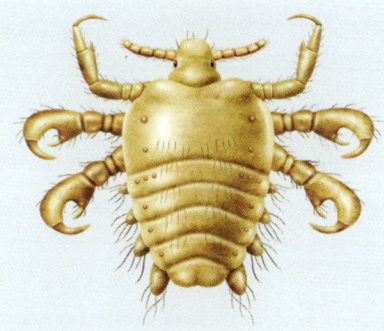

FAMILY Pthiridae

SPECIES IN FAMILY 2

HABITAT Human body

The pubic louse has claws that are highly modified to grasp the flattened human hairs on which it lives. The only other species in the family (*Pthirus gorilla*) lives on gorillas.

To many people, all insects are "bugs."
Strictly speaking, though, the term applies only
to one insect order: the Hemiptera.

HEMIPTERA
TRUE BUGS

The Hemiptera is a huge order of insects, with around 110,000 described species belonging to 104 families. They are often called the "true bugs," and many of the families have the word "bug" in their name—for example, shield bugs, bed bugs, and assassin bugs. Other well-known bugs include pond skaters, water boatmen, froghoppers, and cicadas. Froghoppers (also known as spittlebugs) produce a spitlike secretion on plants, in which their young are protected from predators.

FEEDING AND REPRODUCING The true bugs are a diverse bunch—living on land, in freshwater, and even in the sea—but they all have one feature in common: their mouthparts are stretched out to form a long tube called a rostrum. This allows them to probe deep into plant or animal tissue and to drink sap, blood, or insect fluids. Their feeding method is one reason that many true bugs are pests. Aphids, for example, spread viruses between plants, including staple crops. They have phenomenal reproductive prowess, and for most of the year, aphids do not need to mate to multiply. If the 50 or so offspring of one aphid survive long enough to give birth to their own full complement, and so on throughout a single year, they have the potential to produce a layer of aphids 93 miles (150 km) deep covering the Earth's surface. Fortunately for us, most are lost through predation, inclement weather, and failure to find food.

Not all true bugs are a nuisance. The damsel bugs, for example, feed on aphids and are valuable pest controllers. Some scale insects are used for making dyes, such as cochineal. Even the raucous cicadas have a plus side: they are rich in protein and form part of the human diet in some countries, such as Australia and Mexico.

NABIS CAPSIFORMIS
Pale damsel bug

FAMILY Nabidae

SPECIES IN FAMILY Around 200

HABITAT Various, including gardens, agricultural crops, and forests

These predatory true bugs have enlarged front legs, similar to mantids, to hold their insect prey.

HALOBATES MICANS
Water strider

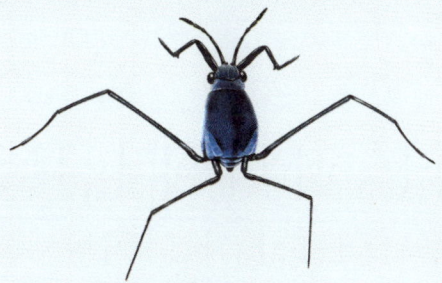

FAMILY Gerridae

SPECIES IN FAMILY Around 860

HABITAT Marine

Halobates micans is one of five *Halobates* species that live on the surface of the open ocean, the only insects that do so.

MAGICICADA SEPTENDECIM
Decim periodical cicada

FAMILY Cicadidae

SPECIES IN FAMILY Around 3,000

HABITAT Woodland, gardens with deciduous trees

Cicadas are known for the loud noise made by adult males, but they actually spend most of their life underground.

HALYOMORPHA HALYS
Brown marmorated stink bug

FAMILY Pentatomidae

SPECIES IN FAMILY Around 4,700

HABITAT Very varied; some species overwinter in buildings

Stink bugs are named after the ability of some species to release a strong smell when threatened.

CORIXA PUNCTATA
Lesser water boatman

FAMILY Corixidae

SPECIES IN FAMILY Around 500

HABITAT Ponds, lakes, slow-moving streams

These aquatic insects were named after their swimming style, which is similar to the action of rowing a boat.

PLATYCOTIS VITTATA
Oak treehopper

FAMILY Membracidae

SPECIES IN FAMILY Around 3,500

HABITAT Very diverse, mostly on trees and shrubs

These bugs have an enlarged plate covering the thorax that often resembles plant thorns.

*More than 6,500 species of thrips have been described
so far. Many are pests, damaging garden plants and field crops
by directly feeding off them and spreading plant viruses.*

THYSANOPTERA
THRIPS

FRANKLINIELLA OCCIDENTALIS
Western flower thrips

FAMILY Thripidae

SPECIES IN FAMILY Around 2,000

HABITAT Various

The Thripidae is the largest family of thrips, specializing in living in narrow spaces, such as within flowers. They are sometimes known as thunderflies, storm flies, or thunderbugs because they often occur in large numbers during stormy summer weather.

Thrips are commonly referred to as fringe-winged or bladder-footed insects. They are tiny and slender (typically less than $\frac{1}{10}$ in/2 mm long) and are tan, brown, blackish-brown, or black in color.

FEEDING Thrips mostly eat plant material. They live in tender foliage, flowers, grass sheaths, twigs, buds, and plant galls (outgrowths). Many species that live in flowers feed on pollen, and some spoil the blooms with their feeding, egg-laying, skin-shedding, and excreting activities. Others inhabit the bark of living or dead trees, rotting or drying grass, and leaf litter. Generally, thrips feed by puncturing the outer layer of the plant tissue and sucking out the cell contents. This causes discoloration, browning, or silvering of leaves. Some species feed on the spores and hyphae of fungi. Still others are predators and feed on other species of thrips, mites, and whiteflies.

CROP FEEDERS Several thrips species are significant crop pests worldwide. Of these, *Thrips tabaci* (known as onion thrips, potato thrips, tobacco thrips, or cotton seedling thrips) is the most common. It attacks a number of major crops, such as those after which it is named, as well as various types of cucumbers, squash, and melons.

*There are currently only two families of snakeflies
(Raphidiidae and Inocelliidae), but fossils reveal much
higher diversity during the Cretaceous and Jurassic periods.*

RAPHIDIOPTERA
SNAKEFLIES

Looking like winged snakes or tiny dinosaurs, snakeflies are closely related to lacewings (p.166) and dobsonflies (p.167). There are about 280 species in two families, and they occur in coniferous and deciduous forests of temperate regions in the northern hemisphere. Larvae need to experience a period of cold weather to induce pupation, so they are often found in mountainous areas.

PREDATORY PUPAE Unlike other insects that show complete metamorphosis (pp.28–29), snakefly pupae can move around and are able to catch prey and feed. Adults are day-flying predators, consuming aphids and mites. They take part in elaborate courtship displays before mating, which involve grooming their partner's legs and antennae. Females use their long ovipositor to lay their eggs in crevices in bark. After they hatch, larvae live under the bark, at the base of trees, amid rocks, or in leaf litter. Like the adults and pupae, larvae are predators and feed on small arthropods.

MORE TO LEARN In the past, many snakeflies lived in tropical regions. It is unclear why the order has become more restricted to temperate regions and much less diverse.

PHAEOSTIGMA MAJOR
Große Kamelhalsfliege

FAMILY Raphidiidae

SPECIES IN FAMILY Around 230

HABITAT Conifers and deciduous trees

The pronotum and mobile head of insects in this order gave rise to the name "snakefly." The German common name of this particular species means "large camel neck fly."

This insect order evolved around 250 million years ago. More than 150,000 species have been identified, but up to 10 times more are yet be discovered. This means Hymenoptera is likely to be the most species-rich insect order, even more so than Coleoptera.

HYMENOPTERA
SAWFLIES, WASPS, BEES, AND ANTS

SAWFLIES The oldest Hymenoptera are the sawflies, known also as horntails or woodwasps. Sawflies, such as the European pine sawfly (*Neodiprion sertifer*), use their sawlike ovipositor to lay their eggs deep within a plant; the larvae then feed on the leaves, buds, and shoots. A small group of sawflies (Orussidae) evolved the ability to exploit meat. They have blind, legless larvae that feast on beetle larvae.

PARASITOIDS True exploitation of prey evolved in the parasitoids (pp.68–69). Many parasitoids are extremely small and hard to identify. They include the smallest insects—the fairyflies (Mymaridae)—which can be less than 0.14 mm long. Parasitoids are important regulators of insect populations, and many species have been exploited as biological control agents of agricultural pests.

HUNTING WASPS Twenty-five million years later, the ovipositor of the sawflies and parasitoids evolved into a sting, giving rise to the hunting wasps. These insects catch their prey, paralyze it with their sting, and place it in a nest, along with an egg. Most hunting wasps tend to be prey specialists. For example, Pompilidae hunt only spiders. Bees are vegetarian wasps that evolved from the Crabronidae family. The majority are solitary, provisioning their brood with pollen.

SOCIAL HYMENOPTERA The Hymenoptera are possibly best known for their social species: bees, wasps, and ants that have evolved the ability to live in a group. Colony sizes of these social insects vary from a few individuals (as in the hover wasp *Liostenogaster flavolineata* of Southeast Asia) to the superorganismal societies of honeybees, vespine wasps, and most ants.

NEODIPRION SERTIFER
European or redheaded pine sawfly

FAMILY Diprionidae

SPECIES IN FAMILY Around 60

HABITAT Conifer forests

Native to Europe and Asia, this species of sawfly was accidentally introduced to North America about 100 years ago, and it has become a pest there as well as in its native range. Similar to many invasive alien species, it can be transported on clothing and in vehicles.

ENCARSIA FORMOSA
Encarsia wasp

FAMILY Aphelinidae

SPECIES IN FAMILY Around 1,200

HABITAT Various, including greenhouses

This parasitic wasp (or parasitoid) is used for the biological control of whiteflies on vegetables in greenhouses across the world. It has been employed for controlling pests since the 1920s.

XYLOCOPA VIRGINICA
Eastern carpenter bee

FAMILY Apidae

SPECIES IN FAMILY Around 6,000

HABITAT Very varied

Carpenter bees are so called because they nest in structural timber as well as in dead wood and cavities in vegetation. Although some of the larger species are seen as pests because they damage timber, they are also important pollinators of plants such as blueberry crops.

FORMICA RUFA
Red wood ant

FAMILY Formicidae

SPECIES IN FAMILY Around 15,000

HABITAT All terrestrial habitats

Formica rufa is one of several similar species commonly known as wood ants. Its huge nests are often conspicuous in mature forests, and they can house more than 100,000 workers. *Formica* species can form supercolonies, and those of *F. yessensis* are known to contain 300 million ants.

The earliest Neuroptera, or net-winged insects,
in the fossil record appeared during the Permian period.
Currently, there are about 6,000 species in 16 families worldwide.

NEUROPTERA
LACEWINGS, ANTLIONS,
AND MANTIDFLIES

LIBELLOIDES
MACARONIUS
Owlfly

FAMILY Ascalaphidae

SPECIES IN FAMILY Around 450

HABITAT Various, widespread

Adult owlflies are predators of flying
insects, and the larvae are ambush
predators. The larvae of some species
disguise themselves with sand, soil,
or plant material.

DELICATE-LOOKING PREDATORS Giant lacewings,
lacewings, antlions, and other Neuroptera are soft-bodied insects
with four membranous wings, which are often delicate in
appearance. This delicacy disguises the voracious nature of the
predatory larvae in most families. These have elongated mandibles,
adapted for piercing and sucking, and many species prey on aphids
and other soft-bodied true bugs. As such, they are used for the
biological control of agricultural pests, and green lacewings
(pp.74–75) are particularly well suited for this purpose.

SPECIALIZED LARVAE Mantidflies, so called because they
look like preying mantids (p.158), have highly specialized larvae,
variously preying within spider egg sacs; as sedentary parasitoids
on bee, wasp, or scarab beetle larvae; or as free-living predators of
small arthropods. Antlion larvae dig conical pits in the ground,
hiding in holes at the center to catch any suitable prey that falls in.

The oldest Megalopteran fossils date to the Early Jurassic.
Today, about 400 species exist within the order, divided into two families:
fishflies and dobsonflies (Corydalidae) and alderflies (Sialidae).

MEGALOPTERA
ALDERFLIES, DOBSONFLIES, AND FISHFLIES

PLEATED WINGS The name Megaloptera means "large wing" and is a reference to the large, clumsy wings of the heavily built adults in this order. Unlike the closely related lacewings, mantidflies, antlions, and snakeflies (p.163), the hind wing has a pleat enabling it to fold when at rest.

SHORT ADULT LIFE Alderfly, dobsonfly, and fishfly adults have strong mandibles adapted for chewing, although numerous species do not eat as adults. Many adult Megaloptera are, therefore, short-lived, surviving from a few hours up to a few weeks. The larvae also have a robust body and strong jaws. They are carnivorous and are aquatic in freshwater, around which the adults can also be found. Larvae develop slowly, taking anywhere from one to five years to mature, at which point they crawl out onto land to pupate in damp soil or under logs. Alderflies, as their name suggests, are found in freshwater habitats where alder trees grow. Fishflies probably got their name from the aquatic lifestyle of their larvae, but why dobsonflies are so called is a mystery.

ACANTHACORYDALIS FRUHSTORFERI

Dobsonfly

FAMILY Corydalidae

SPECIES IN FAMILY Around 320

HABITAT Widespread, with aquatic larvae

Found in Vietnam and China, the *Acanthacorydalis fruhstorferi* species of dobsonfly is the largest aquatic insect in the world, with a recorded wingspan of up to 8½ in (21.6 cm).

The Strepsiptera order is relatively small and not particularly diverse, with only about 600 living species in 10 families.

STREPSIPTERA
TWISTED-WING INSECTS

STYLOPS ATER
Stylops

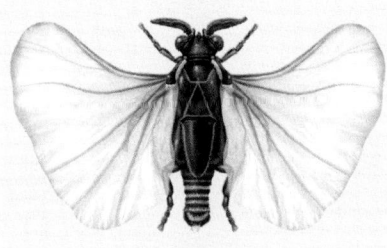

FAMILY Stylopidae

SPECIES IN FAMILY 160

HABITAT Parasites of bees, wasps, flies, and other insects

Special adhesive segments at the end of their feet help the young *Stylops* larvae remain attached to their bee host in flight.

PARASITES Strepsiptera are commonly known as twisted-wing insects, and they are parasites of other insects. They live on hosts across a wide range of other insect orders, including Diptera (true flies, pp.174–175), Hemiptera (true bugs, pp.160–161), Hymenoptera (sawflies, ants, bees, and wasps, pp.164–165), and Orthoptera (grasshoppers, crickets, and bush-crickets, pp.150–151). Their parasitic lifestyle means that twisted-wing insects have complex life cycles. In general, this includes a free-living larval stage. The larvae find a host, often by hitching a ride on an unsuspecting adult insect to its nest, then attack its offspring. The twisted-wing insect larvae develop through several stages within the host, after which the males pupate. Winged adult males then have a brief life outside the host, searching for a mate.

CLEVER FEMALES The females maintain a larvalike appearance, and most of them continue to live within the host throughout their adulthood. Twisted-wing insect females can manipulate the behavior of their host, making it climb up a plant so it is more easily found by flying males, for example. Even though twisted-wing insects are widespread, they are rarely encountered because of their elusive parasitic nature, living most of their lives within their host.

This small order of insects contains about 600 species in 34 genera from nine families. The name Mecoptera means "long-winged," and fossil records date back to around 300 million years ago.

MECOPTERA
SCORPIONFLIES, HANGINGFLIES, AND SNOW FLEAS

CARNIVORES Insects in this order have an important place in the evolutionary tree of insect life, their ancestral lineages having given rise to major groupings of insects. They include scorpionflies, hangingflies, and snow fleas (also known as snow scorpionflies), and they are closely related to fleas (p.172). Mecoptera develop through complete metamorphosis (pp.28–29), and both adult and larval stages are carnivorous. They are predators or scavengers of a variety of dead animals, mostly other insects.

The largest family in the group is the scorpionflies. They do not sting like a scorpion and are, in fact, harmless, but the genitals of males form an upturned tip to their abdomen that resembles a scorpion's sting. As the common name suggests, the snow fleas superficially resemble fleas. They are active during the winter and can be seen walking across snow. The hangingflies could easily be mistaken for craneflies, one of the larger families of the true flies, because of their large, membranous wings and long legs. However, they are different because they have two pairs of wings and no halteres (modified hind wings used to stabilize flight).

Mecoptera generally inhabit moist environments, such as broad-leaf woodlands with plentiful damp leaf litter and moss, where they breed. Courtship behavior is often elaborate, with males offering females nuptial gifts in the form of captured insect prey.

PANORPA COMMUNIS
Common scorpionfly

FAMILY Panorpidae

SPECIES IN FAMILY Around 500

HABITAT Hedgerows and patches of nettles

As is typical of scorpionflies, this species is carnivorous. It feeds not only on dead insects, but also on live aphids and other insects trapped in spider webs.

Beetles are one of the most diverse groups of insects on Earth, with more than 370,000 species named so far by scientists and possibly another million still to be given scientific names.

COLEOPTERA
BEETLES

EVOLUTION Beetles originated more than 300 million years ago, in the Carboniferous, predating dinosaurs by about 100 million years. Some beetle groups evolved in tandem with flowering plants, while others evolved later during the Late Cretaceous—when flowering plants became dominant. Beetles also co-evolved with microbes, and by obtaining genes from bacteria and fungi, they are able to digest plant cell walls and plant sugars. These remarkable life histories probably contribute to the enormous diversity of plant-feeding beetles.

ECOLOGY Beetles are active day and night in both terrestrial and aquatic environments. Their defining feature is a pair of hardened forewings called elytra, which fit tightly over the abdomen to protect the delicate hind wings. They are important herbivores, predators, decomposers, and pollinators, helping in pest control and nutrient cycling, but also causing damage by eating crops and trees. Many beetles use chemicals to defend themselves from predators. For example, blister beetles release a potent toxin called cantharidin, many leaf beetles take up toxic chemicals from the plants they eat to make themselves taste bad, and bombardier beetles produce a fiery spray of hot chemicals when threatened. Longhorn beetles and weevils are masters of chemical communication, using volatile pheromones (chemical gases) to attract mates.

THE FUTURE The sheer number of beetle species offers many opportunities for studying how insects evolve and adapt. However, numerous beetles are threatened by habitat loss, pollution, nighttime lighting, and climate change. Conserving beetle habitats is essential for protecting beetle biodiversity.

CURCULIO PROBOSCIDEUS
Acorn weevil

FAMILY Curculionidae

SPECIES IN FAMILY More than 55,000

HABITAT Widespread

In some weevils, the rostrum (or snout) is not only used for feeding, but also for preparing oviposition sites and placing eggs deep inside plant tissues.

EXOCHOMUS QUADRIPUSTULATUS
Pine ladybug

FAMILY Coccinellidae

SPECIES IN FAMILY Around 6,000

HABITAT Widespread

Many ladybugs (also known as ladybirds) are voracious predators. Some specialize on only a very few insect species, but others prey on hundreds of different ones.

CASSIDA VIRIDIS
Green tortoise beetle

FAMILY Chrysomelidae

SPECIES IN FAMILY Around 3,000

HABITAT Widespread

Tortoise beetles are so called because they pull in their antennae and feet, like tortoises, when threatened.

CICINDELA CAMPESTRIS
Green tiger beetle

FAMILY Carabidae

SPECIES IN FAMILY Around 42,000

HABITAT Widespread

Tiger beetle larvae dig holes in the ground as traps to catch passing insect prey. The adults are thought to be the fastest beetles in the world.

CLYTUS ARIETIS
Wasp beetle

FAMILY Cerambycidae

SPECIES IN FAMILY Around 35,000

HABITAT Forests

Like other cerambycids, wasp beetle larvae feed on dead wood, while the wasp-mimicking adults feed on pollen.

DYTISCUS MARGINALIS
Great diving beetle

FAMILY Dytiscidae

SPECIES IN FAMILY Around 5,000

HABITAT Aquatic

Great diving beetles—up to 2⅓in (6 cm) long as larvae and 1½ in (3.6 cm) as adults—are voracious predators that feed not only on other insects, but also on small fish and newts.

Fleas are a small order of insects that have evolved specializations that make them highly successful parasites of mammals and birds. Worldwide, there are around 2,300 species in 15 families.

SIPHONAPTERA
FLEAS

PULEX IRRITANS
Human flea

FAMILY Pulicidae

SPECIES IN FAMILY Around 170

HABITAT Cosmopolitan parasite

Women are bitten more frequently by fleas than men, indicating that there is a preference for female hormones.

Adult fleas are wingless, and their flattened body allows them to creep between the hairs or feathers of their host. Their very strong and dark cuticle protects them from being removed by the host's grooming activities, and their long hind legs enable them to jump short distances from one host to another. In fact, fleas have amazing jumping abilities, accelerating quickly to leap more than 12 in (30 cm) in less than 0.02 seconds.

CAUSING PROBLEMS Fleas can cause diseases and therefore are a significant public health concern. Their mouthparts are well designed to pierce their host's skin and suck their blood, and a blood meal is usually essential for adult fleas to produce eggs or sperm. Flea bites can irritate the host, leading to skin disease and secondary infections. And fleas themselves can be the host of other organisms, such as the bacterium *Yersinia pestis*, which caused the Black Death plague in Europe in the 14th century. Carried by rat fleas, it probably killed about half of the region's population.

LIFE CYCLE Female fleas can lay up to 50 eggs per day, which then develop through three larval stages to a pupa stage, which can lay dormant for up to five months awaiting favorable conditions. Fleas associated with migrating hoofed mammals and birds, such as the house martin flea (*Ceratophyllus hirundinis*), have one generation per year, whereas a new generation of cat fleas can emerge after only 20 days.

There are about 14,000 species of caddisflies, split into two suborders—Integripalpia and Annulipalpia—based on their feeding behaviors.

TRICHOPTERA
CADDISFLIES

FOOD AND SHELTER The larvae and pupae of caddisflies are found worldwide in a variety of aquatic habitats, including streams, rivers, and ponds. Larvae protect themselves by spinning silk cases or retreats to live in. Their silk often incorporates gravel, twigs, grains of sand, and leaf fragments that help strengthen and camouflage it. Integripalpia larvae live in a transportable case and crawl around in search of food. Annulipalpia larvae live in a fixed retreat that is attached to a stone or plant. They have a varied diet, feeding on plankton, plant fragments, or other small aquatic animals. Larvae can produce a sophisticated net of silk threads to help trap passing food.

LIFE CYCLE Caddisfly larvae pupate in their sealed retreat or case, and pupae have mouthparts to help them escape. They then make their way to the water's surface, where the pupal skin splits and the adult flies away. Adults are nocturnal and usually rest during the day, characteristically holding their hairy wings like a tent over their body.

QUALITY CONTROL The sensitivity of caddisflies to changes in their aquatic habitats means that their presence and abundance are used as bioindicators of water quality. Rivers and streams that have many caddisfly species indicate that the water is clean. The types of species present (in terms of how sensitive they are to habitat changes) can be an indicator of levels of water pollution.

RHYACOPHILA FASCIATA
Sand fly sedge

FAMILY Rhyacophilidae

SPECIES IN FAMILY Around 500

HABITAT Fast-flowing water and widespread

Like most adult caddisflies, this species has subdued patterning that provides camouflage as it rests during the day.

Globally, there are around 170,000 species of fly currently described, although it is estimated there could be as many as 400,000 to 800,000. The earliest fossils date to around 260 million years ago.

DIPTERA
TRUE FLIES

What is a (true) fly? The word "Diptera" comes from the Greek *di* (two) and *ptera* (wing) because, unlike most insects, flies have only one pair of wings—the forewings—that are fully formed for flight. Evolution has reduced the hind wings to tiny, club-shaped appendages called halteres. These unique sensory organs act as in-flight gyroscopes that give the fly stability in the air. In fact, flies are covered in sensory organs. Their bodies are coated in hairs and bristles that transmit signals directly to the brain, helping them make sense of their environment. Fly eyes are generally large, convex, and compound, and in some male flies, the eyes are holoptic: they meet at the top of the head, giving the insect almost 360° vision.

LARGE AND SMALL The lower Diptera include fly families such as biting midges, fungus gnats, mosquitoes, and craneflies. They are long and slender, with extensive and sometimes feathery antennae. The upper Diptera have a chunkier body and shorter antennae and are the species we typically recognize as flies, such as houseflies, blowflies, fruit flies, and hoverflies. Able to live in almost all habitats, from deserts to the depths of the Antarctic, flies are very diverse. The larvae of many species even live underwater. For example, common drone fly (*Eristalis tenax*) larvae live in polluted water such as drainage ditches. They are known as rat-tailed maggots because they have a specialized organ several times their body length called a siphon at their rear end, which they use for breathing. Some flies are predatory, hunting other insects and invertebrates, and many are parasitic, laying their eggs on or inside the body of another animal. Flies are incredibly important for the health of the planet. Among other things, they are brilliant recyclers, eating everything from fungi to rotting plants and dead animals.

BOMBYLIUS MAJOR
Dark-edged bee-fly

FAMILY Bombyliidae

SPECIES IN FAMILY Around 5,000

HABITAT Various, including sandy soil, gardens, parks, heathland, grasslands, woodlands

Although fluffy and cute to look at, bee-flies are parasitoids. They secrete their eggs into a solitary bee nest, and their larvae eat the host larvae and food.

ERIOTHRIX RUFOMACULATA
Red-sided tachinid

FAMILY Tachinidae

SPECIES IN FAMILY Around 10,000

HABITAT Various, including urban, woodlands, grasslands, gardens, parks, coastal

Tachinids are as important to global pollination as bees because they visit many different flowers across wide areas to feed on nectar.

UROPHORA CARDUI
Thistle gall fly

FAMILY Tephritidae

SPECIES IN FAMILY Around 5,000

HABITAT Grasslands, brownfield sites, fallow farmland, marsh

These flies have striking markings on their wings, which they wave and flap to communicate with each other.

HELOPHILUS PENDULUS
Tiger hoverfly

FAMILY Syrphidae

SPECIES IN FAMILY Around 7,000

HABITAT Various, including grasslands, wetlands, woodlands, gardens, parks

Tiger hoverfly larvae develop in still water and are often found in garden ponds.

The term "Lepidoptera" stems from the ancient Greek words lepís (scale) and pteron (wing), after the microscopic scales on the wings of adult moths and butterflies. An astonishing 10 percent of all described species on Earth belong to the Lepidoptera order of insects.

LEPIDOPTERA
BUTTERFLIES AND MOTHS

BUTTERFLY OR MOTH? Most butterflies are diurnal and most moths are nocturnal, but there are exceptions. Adult Hedylidae butterflies are active at night, for example, and burnet moths (Zygaenidae) are sun worshippers. In the field, butterflies can often be distinguished from moths by their club-shaped antennae, whereas moths tend to have feathery or tapering ones. Studies that have explored the evolutionary origins of butterflies found that butterflies form a superfamily (Papilionoidea), which emerged within the moths, meaning butterflies can be regarded as diurnal moths.

WING COLOR Wing patterns help attract mates, give camouflage, or ward off predators. Color is obtained through pigments that are deposited during development or by physical structures that reflect light. The brilliant blue of the morpho butterfly (*Morpho helenor*) is created by ridges on the wing scales that reflect blue light in an iridescent manner due to the angle it hits the wing.

COPING WITH THE CLIMATE In temperate climates, most butterflies and moths produce one or two generations per year, although some—such as the silver Y moth (*Autographa gamma*)—are capable of more. Many species undergo diapause, a state in which development is stalled until the climate is suitable for growth. Pausing development enables individuals to tolerate a range of environments, from mountain peaks to deserts. An alternative way of dealing with unfavorable conditions is to migrate to warmer climates. The painted lady (*Vanessa cardui*), for example, travels some 2,500 miles (4,000 km) in the fall from Europe across the Sahara desert to overwinter in tropical Africa.

TYRIA JACOBAEAE
Cinnabar moth

FAMILY Erebidae

SPECIES IN FAMILY Around 26,000

HABITAT Various, widespread

The larvae of cinnabar moths feed on the ragwort plant, storing its toxic alkaloids. Although these are harmless to the moths, they render the insects toxic to bird predators.

MORPHO HELENOR
Helenor blue morpho

FAMILY Nymphalidae

SPECIES IN FAMILY Around 8,000

HABITAT Various, widespread

Morpho butterflies are some of the largest in the world, with wingspans of up to 8 in (20 cm). Their bright blue color may have several functions, including signaling to potential mates and warning predators.

OPISTHOGRAPTIS LUTEOLATA
Brimstone moth

FAMILY Geometridae

SPECIES IN FAMILY Around 24,000

HABITAT Various

Geometrid larvae are often called loopers, or inchworms, because of their looping way of walking, caused by them lacking several prolegs (unjointed, fleshy appendages) that other families have.

GRIPOSIA APRILINA
Merveille du jour moth

FAMILY Noctuidae

SPECIES IN FAMILY Around 17,000

HABITAT Various, widespread

This family contains some of the most beautiful moth species, but it also includes many major crop pests, such as cutworms and armyworms.

ADSCITA STATICES
Forester moth

FAMILY Zygaenidae

SPECIES IN FAMILY Around 1,000

HABITAT Various, widespread

Unlike most moths, those in the Zygaenidae family are day-flying species. Their bright colors make them some of the easiest to spot.

PAPILIO MACHAON
Old World swallowtail butterfly

FAMILY Papilionidae

SPECIES IN FAMILY Around 550

HABITAT Various, widespread

Swallowtail butterflies are named after the forked-tail-like parts of their wings, which may serve to confuse potential bird predators because of their resemblance to a swallow's tail.

Parasitoid wasp (*Cotesia glomerata*)

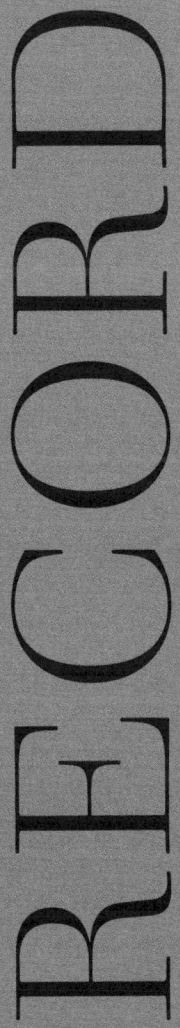

RECORD

CHAPTER VI

HOW DO WE OBSERVE insects and why? How do we know which insect species are declining and which are increasing? Learn how scientific information is gathered, why insects are so important to the ecosystems around them, and how insect populations are monitored.

The general decline in the abundance and diversity of insect life
on Earth is a global concern. Recording and monitoring
insects is crucial to understanding where they are,
what they do, and what they need to survive.

WHAT DO WE OBSERVE AND WHY?

CANARIES IN THE COAL MINE

From 1911 to 1986, miners in the UK used canaries to detect toxic gases before they reached levels that would harm humans. Similarly, insect trends need to be monitored before it is too late. Insects are sensitive to human impacts on the environment, including changes in land use and pollution. They are a key component in ecosystems, and humans depend on them to survive. Monitoring insect numbers is vital to help inform measures to conserve them.

WHAT It is important to understand each insect species and its role in the environment, including its life cycle, where it lives, and what it feeds on. It is also necessary to know how insects interact with each other, because this helps identify their function in an ecosystem.

WHEN Researchers can observe species or communities of insects and take measurements to find out comprehensively what is in a given place at a given time. However, numbers vary dramatically during the year (due to weather, life cycle, food availability, and so on), so a single time point measurement will only reveal so much. This is where monitoring comes in. It can range from measurements made every hour to those made as infrequently as yearly, depending on the species and the questions that need to be answered.

HOW Long-term datasets can reveal trends, such as declines in insect numbers. Samples are typically taken from the natural environment using specialized insect traps, such as light traps that attract night-flying insects. However, insects can also be recorded going about their natural behaviors by observing and counting pollinators when walking a set route, for example. New technologies, such as camera traps, bioacoustics (sound) monitoring, and counting insects passing through lasers, show promise for the noninvasive gathering of information.

Burnet moths
The six-spot burnet moth (*Zygaena filipendulae*, right) is the most common of the burnet moths in the UK. Other species, such as the New Forest burnet moth (*Zygaena viciae*), are very rare. Successful conservation action is being taken, however, supported by monitoring (pp.204–205).

Orange ladybug (*Halyzia sedecimguttata*)
Observations of orange ladybugs that have been collated through community science in the UK have shown how this mildew-feeding ladybug has become more widespread in recent decades. Historically, it was considered to be an ancient woodland specialist, but now it is found in many different habitats, including towns and cities.

Every individual

matters. *Every individual*

has a **role to play**.

Every individual

makes a difference.

~ Jane Goodall, 1994 ~

People have been donating their time in support of insect science for hundreds of years, but in recent decades, there has been a huge increase in the number of volunteers—known as community scientists.

COMMUNITY SCIENCE

Community science is any activity that involves volunteers in scientific research. In early studies, people mostly recorded the occurrence of insect species at a specific place and time. These biological records, accumulated over many decades, are invaluable for answering scientific questions, such as "How quickly are insect distributions altering with climate change?" (pp.200–201).

PROJECT DESIGN Some projects can be relatively straightforward: people can help monitor pollinating insects by submitting their occasional observations of bees, hoverflies, or other insects to recording platforms such as iNaturalist. Other projects have more complex protocols. For example, in the Flower-Insect Timed Count, volunteers count all the different types of insects that visit a patch of flowers over a 10-minute period. Some community scientists design their own research projects, analyzing and sharing results and acting upon their findings.

NEW DEVELOPMENTS Increasingly, smartphone applications and online recording forms are playing an important role in sharing insect observations. Artificial intelligence (AI) can also be used to identify insects submitted through some applications. It is exciting to see the developments in insect monitoring that are linking new technologies—such as AI, environmental DNA, and bioacoustics—with traditional observation techniques to increase the amount of information amassed. These technologies are particularly useful in regions that are difficult to reach, or for insects that are hard to observe by eye. They help provide a better understanding of the amazing diversity of insects and how insects are being affected by human activities.

Why is the descriptive naming of life on Earth so important?
Names help us classify, conserve, and protect insect diversity,
as well as record insect numbers and distribution.

TAXONOMY AND SYSTEMATICS

Taxonomy, or the naming and classification of organisms, is the most fundamental of all biological sciences. Without it, we would not be able to order biological life and understand the relationships, life histories, and functions of insects. The sister to taxonomy is systematics, the study of the diversity of biological organisms and how they are related through evolutionary time. Combined, these sciences help us describe the incredible diversity found in the insect world.

NAMING AND DESCRIBING Carl Linnaeus first proposed the classification system used today in the 10th edition of his *Systema Naturae* (1758–1759). The system is known as binomial nomenclature, in which there is a genus name followed by a species name. For example, the seven-spot ladybug is within the genus *Coccinella* but is a distinct species: *Coccinella septempunctata*. Today, descriptive taxonomy relies on documenting the morphological characteristics of an insect (often using microscopy) to describe a new species. In this way, our understanding of species diversity develops and changes as more insight is gained from ever-improving scientific methods. Since the days of Linnaeus, descriptive methods have evolved thanks to advances in microscopy, including scanning electron microscopy (SEM), in particular. This has enabled taxonomists to reveal more morphological details than was possible with a light microscope.

Insects found in the fossil record, including compression fossils and insects trapped in amber, can also be used to further understanding of evolutionary relationships. These ancient species help us reconstruct insect lineages. Today, combining morphological descriptions with the unique genetic code of a given species is the gold standard in taxonomy and systematics.

Phylogenetic tree
The ancestry of insects is often presented as a phylogenetic tree. To read a tree, we look to the root and follow the branches left to right to the tips, tracking the evolutionary history and relationships between different groups of organisms. The tree depicted here is one hypothesis of how insects are related to one another. As our knowledge grows, these relationships may be contested or change.

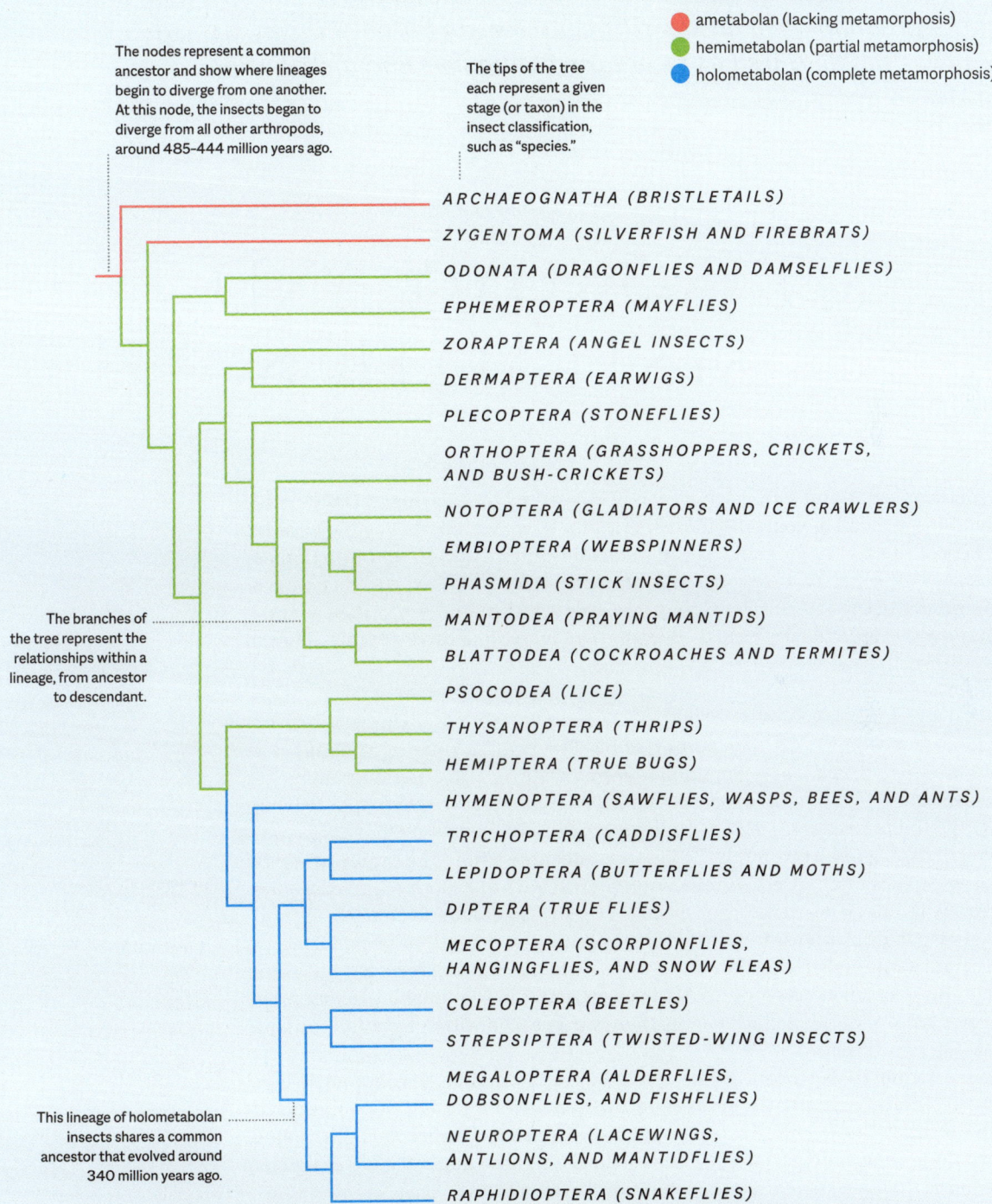

INSECT LIFE CYCLES
- 🔴 ametabolan (lacking metamorphosis)
- 🟢 hemimetabolan (partial metamorphosis)
- 🔵 holometabolan (complete metamorphosis)

The nodes represent a common ancestor and show where lineages begin to diverge from one another. At this node, the insects began to diverge from all other arthropods, around 485–444 million years ago.

The tips of the tree each represent a given stage (or taxon) in the insect classification, such as "species."

The branches of the tree represent the relationships within a lineage, from ancestor to descendant.

This lineage of holometabolan insects shares a common ancestor that evolved around 340 million years ago.

ARCHAEOGNATHA (BRISTLETAILS)

ZYGENTOMA (SILVERFISH AND FIREBRATS)

ODONATA (DRAGONFLIES AND DAMSELFLIES)

EPHEMEROPTERA (MAYFLIES)

ZORAPTERA (ANGEL INSECTS)

DERMAPTERA (EARWIGS)

PLECOPTERA (STONEFLIES)

ORTHOPTERA (GRASSHOPPERS, CRICKETS, AND BUSH-CRICKETS)

NOTOPTERA (GLADIATORS AND ICE CRAWLERS)

EMBIOPTERA (WEBSPINNERS)

PHASMIDA (STICK INSECTS)

MANTODEA (PRAYING MANTIDS)

BLATTODEA (COCKROACHES AND TERMITES)

PSOCODEA (LICE)

THYSANOPTERA (THRIPS)

HEMIPTERA (TRUE BUGS)

HYMENOPTERA (SAWFLIES, WASPS, BEES, AND ANTS)

TRICHOPTERA (CADDISFLIES)

LEPIDOPTERA (BUTTERFLIES AND MOTHS)

DIPTERA (TRUE FLIES)

MECOPTERA (SCORPIONFLIES, HANGINGFLIES, AND SNOW FLEAS)

COLEOPTERA (BEETLES)

STREPSIPTERA (TWISTED-WING INSECTS)

MEGALOPTERA (ALDERFLIES, DOBSONFLIES, AND FISHFLIES)

NEUROPTERA (LACEWINGS, ANTLIONS, AND MANTIDFLIES)

RAPHIDIOPTERA (SNAKEFLIES)

Insects form thriving communities and interconnected webs
that link them to everything from bacteria to birds. Their
interactions create complex ecological networks that determine
the flow of energy and nutrients, and, ultimately, how ecosystems function.

ECOLOGY

DISTRIBUTION Many factors determine whether an insect species can exist in a particular time or place, including environmental conditions, the availability of food, and the presence of predators and diseases. Insects sometimes occupy very specific habitats, and pressures such as climate change and land use change influence where they can be found. As these factors vary over time, the types and numbers of insects within communities also change—as do the interactions among them. Their interactions are vital in many of the ecological processes that we depend upon. Often referred to as ecosystem services, these range from crop pollination and pest control to maintaining healthy soils. Insects can, however, also work against us by performing ecosystem disservices, such as spreading diseases and damaging crops.

NETWORKS IN NATURE Through their interactions with other organisms, insects form ecological networks. By studying these networks, we can begin to understand not only how different organisms are connected to one another and how unique the interactions of some organisms are, but also the consequences of these connections for other organisms that are not directly connected to one another. This includes extinction cascades, in which removing one species from an ecosystem can cause the loss of others that depend on it. Understanding these networks is crucial for mitigating biodiversity loss and ensuring healthy, sustainable ecosystems are maintained.

AVOIDING ONE ANOTHER

Insects can interact without even meeting one another. Many species of ladybug are voracious predators and are widely known for their role in controlling pest insects such as aphids. However, ladybugs are also cannibalistic and will feed on eggs laid by other ladybugs. Adult ladybugs, therefore, avoid laying their eggs on plants where other ladybugs occur. They can detect their presence from the chemicals they leave behind as a trail from their footprints.

Insect interactions
There are numerous important
interactions among species.
Some insects eat plants (yellow)
or pollinate them (blue), while
other species are key predators
(red) or parasitoids (pink).

Key

🟡 herbivory
🔵 pollination
🔴 predation
🟣 parasitism

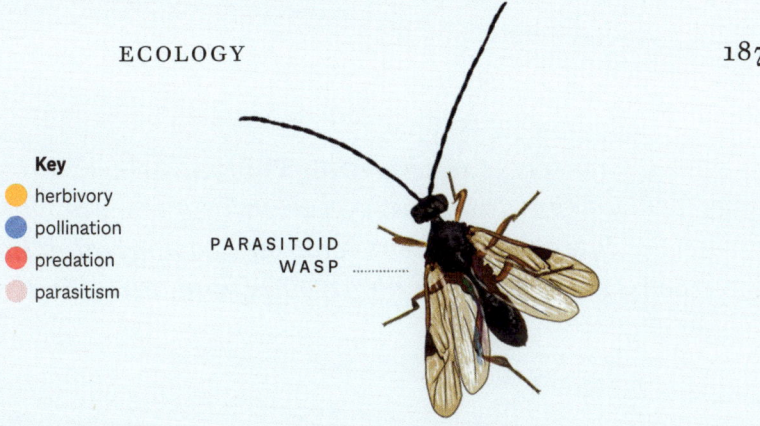

PARASITOID
WASP

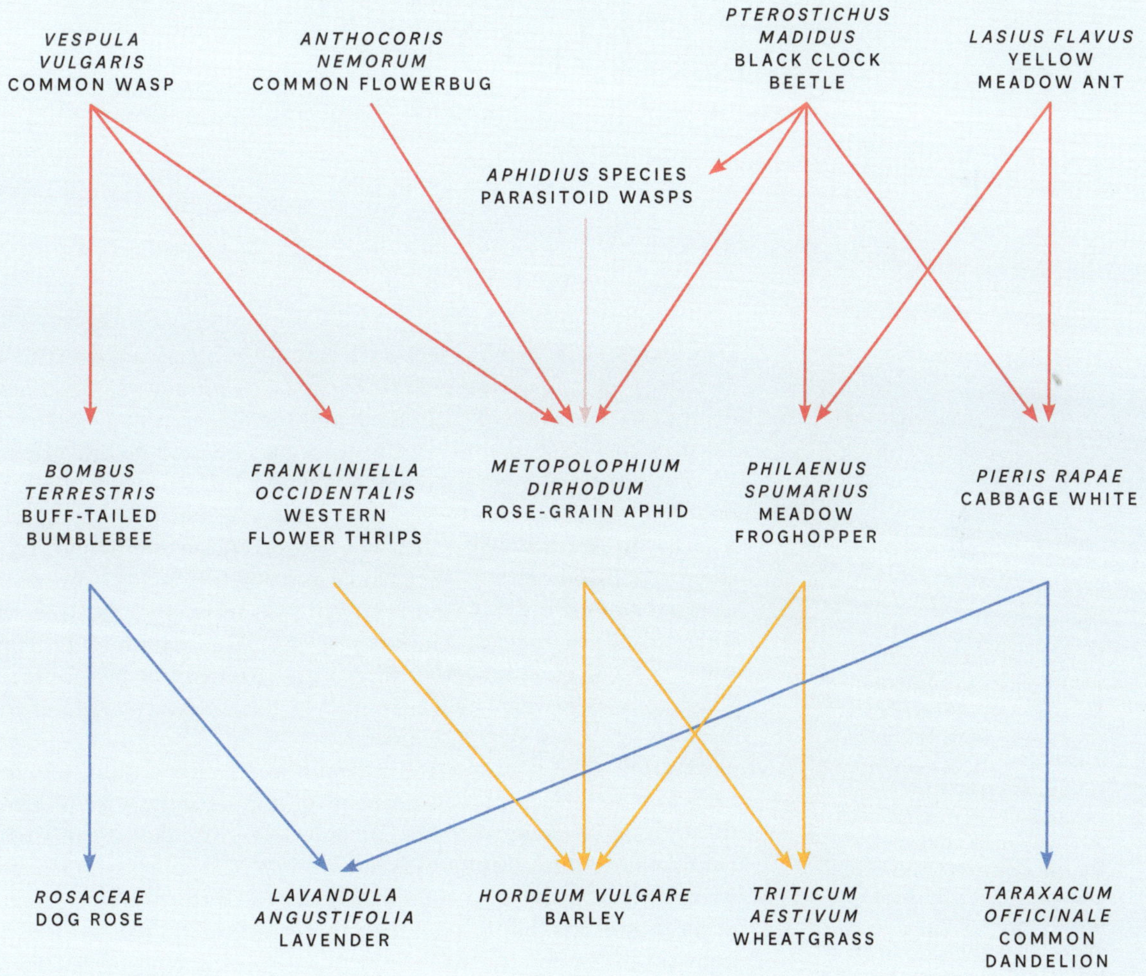

*VESPULA
VULGARIS*
COMMON WASP

*ANTHOCORIS
NEMORUM*
COMMON FLOWERBUG

*PTEROSTICHUS
MADIDUS*
BLACK CLOCK
BEETLE

LASIUS FLAVUS
YELLOW
MEADOW ANT

APHIDIUS SPECIES
PARASITOID WASPS

*BOMBUS
TERRESTRIS*
BUFF-TAILED
BUMBLEBEE

*FRANKLINIELLA
OCCIDENTALIS*
WESTERN
FLOWER THRIPS

*METOPOLOPHIUM
DIRHODUM*
ROSE-GRAIN APHID

*PHILAENUS
SPUMARIUS*
MEADOW
FROGHOPPER

PIERIS RAPAE
CABBAGE WHITE

ROSACEAE
DOG ROSE

*LAVANDULA
ANGUSTIFOLIA*
LAVENDER

HORDEUM VULGARE
BARLEY

*TRITICUM
AESTIVUM*
WHEATGRASS

*TARAXACUM
OFFICINALE*
COMMON
DANDELION

The diversity and abundance of insects with the potential to become crop pests is always changing. Monitoring them is a challenge, but it helps farmers and growers make effective management decisions.

MONITORING CROP PESTS

MAKING DECISIONS The methods used to monitor crops pests change depending on the species and the life stage being assessed. Monitoring several species across a farm is, therefore, time-consuming and often expensive. At the start of each season, farmers select the most important insects to monitor and the methods to use. Not all pests live in crops during their whole lifetime, and many move across large areas of the countryside in different ways from year to year. These migrations of occasional pests can make it difficult for farmers to know if, when, and where to monitor.

TECHNIQUES The simplest monitoring technique is direct observation. This starts with a low-intensity, wide-scale approach, used to observe the first appearance and then overall spread of infestations. It involves regular visits to crops and surrounding habitats across the farm and consultations with neighbors or insect monitoring networks. For some insects, monitoring equipment, such as water traps, may be used to get a clearer and quicker view of activity. Recorded observations of insect numbers can also be employed in software applications, which use models to predict the risk of crop damage. This helps put observations into the context of farm management decisions.

PEST MONITORING NETWORKS

These networks are useful because they share information about sightings of specific insects and can provide an early warning of potential risks to crops. Every year since 1964, the Rothamsted Insect Survey, for example, has collected observations of aphids and other insects at fixed locations across the UK. These data are used to alert growers to the start of aphid migrations each year, enabling them to better target in-field monitoring and management.

ANALYZING ENVIRONMENTAL DNA (eDNA) TO MONITOR BIODIVERSITY

SAMPLE FIELD

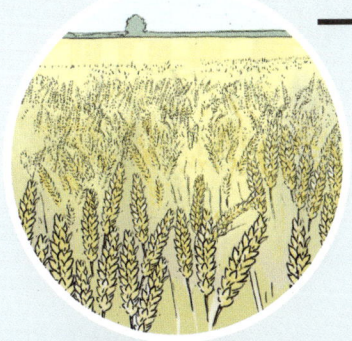

1.

Collecting: DNA is collected from the environment instead of from the organism itself. Typically, it is collected from water, soil, or air, but it could also be from skin, gut contents, or fecal matter.

2.

Processing: This depends on the substrate. For water samples, the water is usually filtered. Other samples may be crushed and suspended in distilled water.

3.

Extracting DNA: DNA extraction is carried out using a specialty kit or standard protocols for invertebrates.

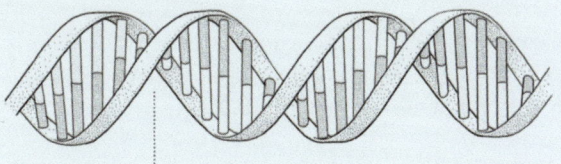

DNA DOUBLE HELIX

4.

Amplifying DNA: To amplify the DNA via polymerase chain reaction, primers are used that either bind to DNA sequences of one species (traditional barcoding) or many species (metabarcoding).

5.

Sequencing: Samples are run on a genetic sequencer to read the DNA sequences. Next-generation sequencing is used for metabarcoding.

6.

Bioinformatics: The DNA sequences generated are compared to a database of known DNA sequences to identify which species are present in the environment from which the samples were taken. This genetic data can be used to inform the monitoring of insect populations, diversity, and distribution without the need to collect live insects.

*... it is important simply to make **welfare** impacts on **insects** salient in discussions of **conservation practices and policies.***

~ Meghan Barrett, Bob Fischer, and Stephen Buchmann, 2023 ~

There are many different contexts in which it is critical to consider the values, principles, and conflicting expectations that influence our choices. Such ethical inquiry, asking questions about what we ought to do and why, is highly relevant to the study of insects.

WELFARE AND ETHICS

Ethics in the study of insects touches on many areas: how research is gathered, animal ownership and management practices, environmental stewardship, medical research and disease management, technological innovation, and more. Two particular areas of interest are fieldwork conservation ethics and animal use ethics.

FIELDWORK CONSERVATION ETHICS Entomologists try to reduce their impact on the environment when mapping the distribution of insects in the field or conducting research. They make the best use of specimens collected and reduce the need to collect more in a variety of ways, by storing bycatch (nontarget insects that get caught in a trap) for future study and making use of existing museum specimens, for example. Entomologists are also developing sampling methods that do not kill insects. These can include photography, recording soundscapes, and using molecular tools such as eDNA (pp.188–189).

ANIMAL USE ETHICS Whether insects are sentient or can feel pain is still an open question. However, even skeptics have recommended applying a version of the "precautionary principle" to using insects within research, whereby steps are taken to reduce harm, in case it can occur. This had led to calls for more consideration of best practice in both laboratory and field research. Some have recommended applying the "three Rs" framework in entomology: insects are "replaced" in studies where possible, the number of insects used is "reduced," and research practices are "refined" to minimize harm. The first scientific guidelines for promoting insect welfare in research were produced in 2023, outlining nutritional, behavioral, environmental, and health aspects of insect welfare.

Buff-tailed bumblebee (Bombus terrestris)

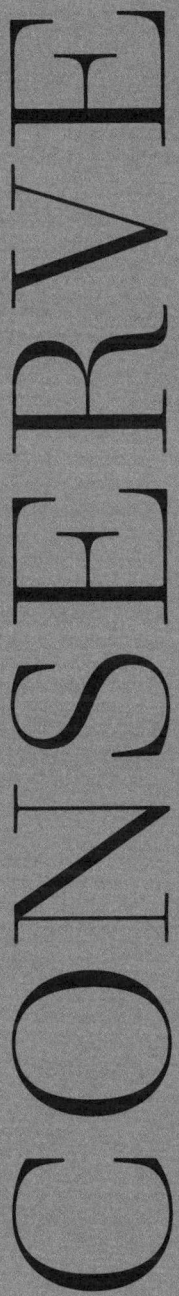

CONSERVE

CHAPTER VII

AS INSECT HABITATS are changing, biodiversity is threatened and insect numbers are declining. Are insects able to adapt to their new environments, and quickly enough? Find out how some insects are changing their ranges as the climate warms, and explore the ways in which we can restore urban and natural habitats to create a more optimistic future for insects.

*If we were to **wipe out** insects alone on this planet ... the **rest of life** and humanity with it would **mostly disappear** from the land. Within a **few months**.*

~ E. O. Wilson, 2007 ~

It is unclear how severe or widespread insect declines are, but there are things that can be done to stop them from happening and to make sure that insects thrive.

INSECT DECLINE

MEASURING DECLINES The global extent of insect decline is not known, because it is very hard to get reliable measures of long-term trends. Knowledge of insects is dominated by a small number of well-studied insects that are economically important, including those that are crop pests, such as grasshoppers, and those that are beneficial, such as honeybees (*Apis mellifera*). There are initiatives to get people recording insects through community science (pp.182–183), but these are currently limited to relatively few insect groups in relatively few parts of the world. Insect numbers can fluctuate considerably and rapidly, mostly because of weather and seasonal changes or human activities. Spraying pesticides, cutting down forests, draining wetlands, and polluting waterways, for example, harm insects and cause their populations to decline. Data need to be collected consistently over decades to calculate if a population is in long-term decline and the rate at which it is doing so. Unfortunately, this information does not exist for most insects.

WHAT CAN BE DONE? The good news is that insect declines can be halted and insect populations restored. There are simple ways that we can help save insects every day, such as not using pesticide sprays, growing trees and flowering plants in gardens and pots around the home, and not squashing bugs. Gardens can be places that both insects and people enjoy (pp.224–225). Changing the ways in which we use land can also avoid the unnecessary loss of insects. We can farm in more nature-friendly ways by having wildflower strips along arable field margins (pp.206–207), make space for nature when we build new houses, and protect more natural habitats. Urgent action is needed to safeguard threatened insects and to make sure that the importance of insects is a key part of conservation initiatives.

**Buff-tailed bumblebee
(*Bombus terrestris*)**
Some insects are active in winter, but warmer winters make them more active, and it is vital that they can find food. The buff-tailed bumblebee searches for nectar in winter to feed her offspring.

**Diamondback moth
(*Plutella xylostella*)**
In Northern Europe, the migratory diamondback moth can cause considerable damage to brassica crops, such as cabbage, broccoli, and cauliflower, and the moth is likely to get more common as a consequence of climate warming.

*Climate change is causing problems for
some insects but benefiting others, including
some that cause us harm.*

CLIMATE CHANGE

REGULATING BODY TEMPERATURE The number of extreme cold and hot weather events is on the rise. Insects are very vulnerable to changes in temperature because they are cold-blooded, so they have very little opportunity to regulate their body temperature above or below that of their surroundings. This means that the air temperature influences all aspects of what they do, including growing, reproducing, and surviving. Climate change is, therefore, having major effects on insects and their interactions with people: altering insect distributions (pp.200–201), raising the risk of invasive insects (pp.198–199), and increasing the occurrences of diseases that are transmitted by insects (pp.96–97).

SURVIVING WINTER In temperate climates, such as the US and Europe, many insects survive cold conditions by becoming dormant and suspending their development. Insects rely on environmental cues to let them know when to do this, and the shorter days and cooler temperatures of fall trigger the onset of winter dormancy. However, climate change is leading to warmer fall and winter temperatures, which are, in turn, disrupting the cues and affecting the timing of dormancy. Many insects are now remaining active throughout winter, but they are not equipped to survive the extremely cold spells.

Warmer winters can boost insect survival, and being active in winter can bring benefits, such as earlier reproduction in spring and faster increases in numbers in summer. Insects that usually migrate to warmer countries can stay put if winters are mild. However, warmer winters are often bad news for farmers, because some insects that cause extensive damage to crops, such as the migratory diamondback moth (*Plutella xylostella*), flourish in these conditions.

The global estimate of the number of alien insect species is more than 10,000, and a high proportion of these have adverse impacts on biodiversity and ecosystems.

ALIEN INSECTS

Alien species are those that have been moved by people from one part of the world to another where they would not naturally occur. Invasive alien species are a subset of alien species that are known to cause damage to ecosystems, economies, and well-being. Some go unnoticed or unreported, but there are many well-known examples, particularly among beetles (pp.170–171); true flies (pp.174–175); butterflies and moths (pp.176–177); and wasps, bees, and ants (pp.164–165).

TROUBLESOME SPECIES Invasive alien insects cause adverse impacts in many ways. Some are voracious predators and generate dramatic declines of other insects. For example, the harlequin ladybug (*Harmonia axyridis*) was introduced to many countries as a biological control agent of pest insects such as aphids, but it spread unintentionally and caused declines in native ladybugs by outcompeting them for food and even eating their eggs and larvae. Other insects are vectors of disease, such as mosquitoes, which not only transmit pathogens to humans, but also to wildlife. Particularly troublesome are invasive alien ants, which can change the function of entire ecosystems (see opposite).

HITCHHIKERS Alien insects are mostly transported unintentionally, for example, as stowaways on traded goods such as ornamental plants. However, some have been introduced intentionally by people to assist with pollinating plants. The buff-tailed bumblebee (*Bombus terrestris*) was introduced to South America from Europe and has been linked to declines in the only native Patagonian bumblebee, *Bombus dahlbomii*, the world's largest bee. The European bees may be superb pollinators, but they are also infected with parasites that kill the native species.

FUTURE ALIENS
Climate change is likely to increase the problem of invasive alien insects in the future. Many more insect species may be transported from invaded ranges by extreme weather events, and others may spread and thrive as the climate warms (pp.196–197) and native species decline. Reducing the effects of climate change will be important for mitigating the impacts of invasive alien species and helping protect biodiversity.

IMPACT OF
INVASIVE ALIEN ANTS

1.

Invasive alien ants, such as the Argentine ant (*Linepithema humile*), are among the world's most invasive alien insects. They are transported in many different ways, including by tourists and international trade.

2.

Ants' nests alter the physical and chemical properties of the soil, which can have a domino effect on other soil-dwelling species if they are unable to live in these highly modified places.

3.

Predatory ants can also alter the abundance and diversity of other animals (Argentine ants have voracious feeding habits and outcompete native insects), which in turn can reduce the food available for insect predators.

4.

Through predation and competition, invasive alien ants adversely impact insects that are involved in seed dispersal and pollination and, as such, disrupt these key ecosystem processes.

5.

Climate change may provide environmental conditions that enable invasive alien ants to thrive in new regions.

6.

Argentine ants use sophisticated chemical communication to limit the size of their colony within their natural range. However, in regions where they have been introduced, this communication system has broken down, and these ants form enormous "super colonies" that can be very damaging if they displace native species.

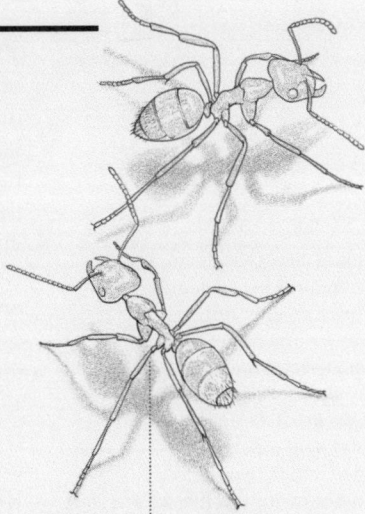

ARGENTINE ANTS
(*LINEPITHEMA HUMILE*)

INSECTS ADAPTING

As well as changing their range, insects in a warming world may alter their behavior and emerge earlier in the year. The flight period of the small blue butterfly (*Cupido minimus*) has advanced by about five days per decade in the UK. Earlier flight periods can be harmful for some species, though, because they may cause timing mismatches with other species in their ecosystem, for example.

Comma butterfly
(*Polygonia c-album*)
This leaflike butterfly camouflages itself well in the woodlands it occupies. It has been tracked moving to more northerly habitats since the 1970s.

*Global climates are changing, and insects are responding.
Species are shifting their ranges to track the changing
climates, resulting in them moving uphill or toward
the North or South Poles to reach cooler locations.*

RANGE CHANGES

A study of butterflies in Northern Europe provided some of the first evidence of the ecological impact of human-caused climate change. It revealed that more than half of the species had shifted their ranges northward by between 35 and 22, or 150 miles (240 km), during the 20th century. This "fingerprint" of the impacts of climate change is evident in many insect groups, including dragonflies, bees, and butterflies and moths.

SPREADING THEIR WINGS In the UK, many species are spreading northward, although the rates at which they are doing so vary considerably. Butterflies are popular with the public, and volunteers provide millions of records for scientists to analyze. This information reveals the most successful butterfly, in terms of responding to warmer temperatures, in the UK is the comma butterfly (*Polygonia c-album*), which is spreading northward at a rate of more than 6 miles (10 km) per year. The speckled wood butterfly (*Pararge aegeria*) is also expanding its range, but rates vary according to the amount of woodland habitat present, which it needs to breed. Expansion is a lot slower where woodlands are absent or fragmented. Climate change is also causing some ranges to diminish: the mountain ringlet butterfly (*Erebia epiphron*) prefers cooler montane locations and has retreated from about a third of its UK sites due to rising temperatures.

These rates of spreading northward are far more rapid in Finland, where the range of the silver-washed fritillary (*Argynnis paphia*) has shifted by more than 186 miles (300 km) in 10 years. In tropical regions, insects are moving uphill instead of toward the poles. Moths on Mount Kinabalu in Borneo, the highest mountain in Southeast Asia, have moved on average 220 ft (67 m) uphill over four decades but have lagged behind climate change. (A shift of about double that observed would be required to track the changing climate.)

Urban environments are often seen as unwelcoming places for insects, but people and insects can coexist and thrive in the "urban jungle" if green spaces are managed with insects in mind.

MANAGING URBAN HABITATS

Individual urban sites—such as gardens, parks, and roadside shoulders—may be small, but they are important insect habitats. Together, they form large networks for insects and connect urban sites with nature reserves and the wider countryside.

INSECT-FRIENDLY PLANTING Gardens, parks, and other green spaces can be incredibly valuable for insects if they are planted to provide food and shelter for them. A diverse array of plants will support a diverse group of insects, so a range of trees, shrubs, grasses, and herbaceous plants should be planted, with different shapes and structures. Native plants are a priority to provide food for specialist insect herbivores. However, many generalist insects will feed on non-native plants. For example, non-native fuchsia is adored by caterpillars of elephant hawkmoths (*Deilephila elpenor*) in the UK, and alien plants such as the butterfly bush (*Buddleja*) provide nectar for urban insects in Europe and are especially valuable if they extend the flowering period. Leaving piles of logs and mounds of leaf litter will also help insects that depend on eating rotting matter.

HOMES AND SHELTERS Urban ponds, even tiny ones, can provide habitats for aquatic insects, including dragonfly larvae and pond skaters. Bug hotels (artificial structures), such as those with tubes for solitary bees or overwintering slots for butterflies, not only provide homes and shelter for urban insects, but also inspire people to watch and admire insect visitors. Night-flying insects and their larvae can be affected by light pollution from outdoor lighting and street lamps, but reducing intensity and beam extent using warmer hues and ensuring lights are turned off whenever possible can help minimize harm.

Attracted to the light
Artificial light can draw moths and other insects away from their natural nocturnal behaviors, such as feeding, mating, and pollinating plants.

Insects occur in vast numbers, so, for conservation purposes, it makes sense to focus on restoring natural habitats instead of saving individuals. If the conditions are right, insect populations can recover more quickly than most other animal groups.

RESTORING NATURAL HABITATS

REVERSING HABITAT LOSS There is a very long history of people altering natural habitats to produce food and timber and to build housing, which has led to insect declines. There is now a focus on restoring these habitats and the insects they support by rewetting drained fens and bogs and reintroducing traditional grazing practices, for example. Fenland areas are being restored in eastern England, where previously about 99 percent of this important habitat had been lost. In 2001, a 50–100-year plan was put in place to restore about 14 sq miles (36 sq km) of wild fen to attract the rare insects that occur there, such as the tansy beetle (*Chrysolina graminis*). Habitat restoration has also focused on bringing back traditional methods of mowing, collecting wood, and livestock grazing. A 1,480-acre (600-ha) site in the White Carpathian mountains (on the border of Slovakia and the Czech Republic) is being restored in this way for the benefit of insects, such as the stag beetle (*Lucanus cervus*), clouded Apollo butterfly (*Parnassius mnemosyne*), and Jersey tiger moth (*Euplagia quadripunctaria*).

BACK FROM THE BRINK The New Forest burnet moth (*Zygaena viciae*) occurs in Europe and Scandinavia, but it has declined in the UK because of the overgrazing of its habitat. It now occurs at a single site, in the west of Scotland. By 1990, only about 20 adult moths remained. Fencing was put up to exclude sheep, thus allowing the reestablishment of the caterpillars' main food plant: meadow vetchling (*Lathyrus pratensis*). The moth population has dramatically increased, and 8,500 to 10,200 individuals were estimated to be present by 2003.

**Large heath butterfly
(*Coenonympha tullia*)**
This species has returned to restored lowland peatland habitats in several parts of the UK, which had been drained during the Industrial Revolution in the 19th century.

Stag beetle (*Lucanus cervus*)
Stag beetles require habitats rich in decaying wood (pp.110–111), where their larvae live for three to four years (pp.30–31). Habitat restoration involves ensuring a continuity of suitable habitat to facilitate the dispersal of adult beetles.

**Scarce chaser dragonfly
(*Libellula fulva*)**
This threatened insect has benefited
from rewilding practices that have
helped restore and improve its
wetland habitat in places that were
previously farmland.

By letting nature take over, we can provide places where insects are able to thrive. These wilder areas can support a wider range of insect species and more abundant insect communities.

REWILDING

RESTORING ECOSYSTEMS Conservation practices to restore natural habitats can breathe new insect life into ecosystems. The term "rewilding" refers to restoring biodiverse landscapes by encouraging trees and shrubs to grow back (including those often seen as weeds) and reintroducing species that play important roles in healthy ecosystems. The Knepp Estate in West Sussex, England, was formerly a dairy and arable farm, but since 2000, it has been rewilded. Dairy cattle have been replaced with free-roaming livestock, and the landscape is being allowed to revert to scrub and woodlands.

THE BUTTERFLIES ARE BACK Caterpillars of the purple emperor butterfly (*Apatura iris*) feed on sallow leaves, so the natural regeneration of sallow scrub at the Knepp Estate has provided ideal conditions for their return. A total of 388 individuals were recorded on a single summer's day in 2018, which was twice as many as the previous year and a UK record. This demonstrates the effectiveness of rewilding and how quickly insects can recover. Many other insect species are now flourishing at the Knepp Estate, including the violet dor beetle (*Geotrupes mutator*) and the scarce chaser dragonfly (*Libellula fulva*).

The Étang de Cousseau nature reserve in southwestern France, where dung beetles (*Scarabaeus laticollis*) had been absent since 1965, has also been rewilded. Around 60 dung beetles were reintroduced in 2023, collected from a site near Montpellier, France, and then transported to the reserve and released into an open dune meadow. The intention is that the dung beetles will create nutrient-rich soil—by transporting rolled balls of cattle dung into underground chambers—and improve the soil health in the nature reserve.

The eye-catching large blue butterfly is making a comeback.
It has been successfully reintroduced in the UK, where it is now
thriving thanks to targeted conservation action. This brings
hope that it can flourish elsewhere.

SAVING A SPECIES
FROM EXTINCTION

THE BIGGER PICTURE
Many other insect species
also declined because of
reductions in grazing and
grasslands becoming
overgrown. Restoring
the grasslands for the
reintroduction of large blue
butterflies has, therefore,
benefited other species.
There have been dramatic
increases in the numbers of
many endangered plants and
insects, including the rugged
oil beetle (*Meloe rugosus*),
pearl-bordered fritillary
(*Boloria euphrosyne*), and fly
orchid (*Ophrys insectifera*).

LIFE STORY The large blue butterfly (*Phengaris arion*) has a
rather unconventional life cycle, which begins when the female lays its
eggs on wild thyme and marjoram flowers. After feeding, the resulting
caterpillar drops to the ground to await discovery by a *Myrmica* ant. A
bizarre "deception" follows. The caterpillar releases a scent that
convinces the ant that it is one of its own offspring. It also puffs up its
body to appear more like an ant grub. The ant takes the caterpillar
underground, where it lives among the ant grubs. There, the
caterpillar becomes a carnivore, eating ant grubs in the nest, before
pupating. The following summer, the adult emerges from the ant
nest, inflates its wings, looks for a mate, and the life cycle is complete.

MYSTERIOUS DECLINE The large blue had always been rare in
the UK, but by the mid-20th century, it was declining rapidly and
nobody knew why. In 1963, the Joint Committee for the Conservation
of the Large Blue Butterfly was formed, with the aim of saving this
iconic species, but by 1970, only two small populations remained in
the UK. Research scientist Jeremy Thomas was given the task of
solving the conservation mystery. He discovered that although five
species of *Myrmica* ants will "adopt" large blue caterpillars, the
insects only survive with *Myrmica sabuleti*, a heat-loving ant that
thrives only on grassland sites that are regularly grazed by livestock.

Unfortunately, these findings came too late to save the large blue. The butterfly population collapsed because changes in livestock farming had led to grasslands becoming overgrown. In turn, this led to declines in red ants and too few ant nests to support butterfly populations. In 1979, the large blue was declared extinct in the UK.

CONSERVATION SUCCESS Once the grassland sites were regularly grazed, the numbers of *Myrmica sabuleti* ants quickly increased. Large blue butterflies were collected from donor populations in Sweden and successfully reintroduced at a site in Dartmoor in 1983 and Somerset in 1992, then to other sites around the UK. Now, UK populations are some of the largest known in Europe, and large blue butterflies are a common sight in summer.

Large blue butterfly (*Phengaris arion*)
Females lay their eggs on the flower buds of thyme and marjoram, which thrive in well-grazed grassland habitats. Here, the short turf is home to nests of *Myrmica* red ants, the unwitting host of the butterfly's parasitic larvae.

Increasingly, we are recognizing the importance of insects on Earth and the harms and benefits they bring. The future will present new challenges and opportunities for insects, which will, in turn, affect people's lives.

WHAT IS THE FUTURE FOR INSECTS?

Insects are cold-blooded, so temperature affects just about everything they do, including growing and reproducing. Warmer temperatures are causing some species to shift their ranges (pp.200–201) and alter their seasonal activities, emerging earlier in spring and being more active in warmer winters. Plant-eating insects suffer when their host plants dry up during droughts or get washed away during floods, and climate projections are for more extreme temperature and rainfall events. Which insects will be affected and how is not well understood. But many other organisms depend on insects, so any future changes will have domino effects, for example, on insect-eating birds and insect-pollinated plants.

WINNERS AND LOSERS Some species benefit from environmental changes, while others decline. Insects that are good fliers and generalists tend to cope better and their numbers are increasing, such as the highly mobile comma butterfly (*Polygonia c-album*), which feeds on a range of common plants including nettles. Insects that are poor fliers and highly specialized tend to be in decline, such as the Duke of Burgundy butterfly (*Hamearis lucina*), whose caterpillars feed only on primrose and cowslips. It is, however, difficult to gauge which species are winning and which are losing, and not all species are increasing or decreasing in all the places they occur. Range shifts in insect vectors of disease are bringing new diseases from tropical regions to cooler ones, and climate change is leading to more crop pests and more arrivals of non-native invasive species. And many insect "winners" from warming temperatures are causing economic and health problems for people. One example is the oak processionary moth (*Thaumetopoea*

processionea), which is a pest of oak trees and whose caterpillar hairs can irritate human skin and eyes. New technologies, such as artificial intelligence to identify insects and bioacoustics to survey them, may help people better understand insect populations, especially in regions that are poorly surveyed.

HOW CAN WE HELP? We can support insect science by being a member of an insect society or conservation organization. By volunteering as a community scientist (pp.182–183), we can collect data on the insects we see around us and upload the information to datasets that can be analyzed to understand long-term trends. Such information is used by governments to help meet their commitments to the Convention on Biological Diversity to halt biodiversity decline. We can also help insects by making their local environment insect friendly: by leaving wild areas in gardens and parks where insects can thrive, choosing garden plants that are good nectar sources, avoiding using chemical pesticides and fertilizers, and spreading the word about the importance of insects.

*If all these **challenges** can be addressed ... **entomology** will have a diverse, vibrant, and **influential future** ...*

~ Sarah Luke et al., 2023 ~

Japanese rhinoceros beetle (*Allomyrina dichotoma*)

CHAPTER VIII

**PEOPLE HAVE BEEN
FASCINATED** by insects for millennia.
Find out how insects have made their way
into literature, art, and museum collections,
as well as onto our dinner plates and into our
homes as pets.

With more than 300 million insect specimens from the 18th century to the present day, the Natural History Museum in London contains one of the world's largest collections of insects. Such museums not only provide information to the public, but also help place current insect trends in a historical perspective.

MUSEUM COLLECTIONS

WHAT MIGHT WE FIND IN A MUSEUM? Open the drawers behind the scenes in a natural history museum and insects are carefully laid out in ordered rows. They have been preserved either by drying them out and pinning them to display their characteristic features or by pickling them in preserving liquid. In a predigital world, this was the only way in which information about insects could be collected and stored. Today, digital technology allows photographic records to be made instead, and views about catching and killing insects for study have changed, especially for insects that can be identified alive, such as ladybugs, dragonflies, butterflies, and large moths. As a result, there has been a recent increase in the number of digital collections made available to the public.

WHAT INFORMATION CAN INSECT SPECIMENS GIVE? Modern technology allows genetic information to be obtained from insect specimens, and measuring their size and shape also provides data on how insects vary and change over time. Specimens are labeled with details about where and when they were collected, but some of these can be hard to interpret if they are written in copperplate script or if historical place names have changed. Sometimes, other things, such as pollen or parasites, are preserved along with the insect, thus providing information on historic insect interactions within ecosystems.

UNDERSTANDING BIODIVERSITY CHANGE Museum specimens and museum data have both facilitated understanding in this field. For example, comparing British swallowtail butterfly specimens from different locations revealed shrinkage in wing size associated with the drainage of their fenland habitats. Measuring pollen loads on bee specimens showed that loss of host plants was a major driver of decline in the Netherlands. And resurveying geometrid moths on a mountain in Borneo, Malaysia, revealed uphill range shifts (pp.200–201) and that present-day communities are made up of smaller moths arriving from the lowlands.

Scarce chasers (*Libellula fulva*)
These dragonfly specimens have been dried, pinned, and labeled. From top to bottom, labels typically describe type; date, location, and collector's name; species name and person who identified it; name of collection; other information such as previous illustration or QR code.

Insects are a rich source of food, and humans have a long history of eating them. Today, about 2,000 species are consumed regularly by around 2 billion people in more than 110 countries.

EATING INSECTS

PREPARING INSECTS

The landibe (*Borocera cajani*) is a delicacy in Madagascar, where the moth's pupa is eaten. The pupa is removed from the cocoon and placed in boiling water before being fried or cooked in ashes.

In most of Southern Africa, from Zambia southward, the mopane worm (*Gonimbrasia belina*) is an important species of edible caterpillar. To cook, the contents of the gut are removed before the caterpillars are boiled and seasoned. They are then fried and eaten with maize porridge, but they may also be skewered and roasted over a fire.

With a similar protein content to conventional meat, edible insects may become a significant part of the human diet in many parts of the world as the demand for food increases. Today, the most commonly eaten insects are crickets, moth caterpillars, grasshoppers, mealworms, and black soldier flies (*Hermetia illucens*). Caterpillars, such as those of the pallid emperor moth (*Cirina forda*), are a key food source in much of Central, Western, and Southern Africa. They are rich in protein and are particularly important when other food sources are low. Sadly, overcollection and habitat loss have led to many moth species becoming scarce.

LABOR-INTENSIVE CROP Blotched emperor caterpillars (*Lobobunaea phaedusa*) are difficult to find because they usually live alone and are well camouflaged among the leaves. They can, however, be found by searching for their droppings below trees. In the Democratic Republic of the Congo, African star apple moth caterpillars (*Achaea catocaloides*) are collected in large quantities and sold as food in markets. They are mainly collected by children, who make loud calls as they work. Caterpillars jerk their body in reaction to sound, which makes them easy to spot.

*Insects for **human consumption** and as feedstock for livestock and fish could **contribute to food security** and be part of the solution to the **meat crisis**.*

~ Arnold van Huis, 2013 ~

With an increasing global population and more demand for animal protein,
there is a pressing need to find alternative and sustainable sources
of livestock feed. Insects are a staple food source for many
animal groups and have emerged as a promising option.

LIVESTOCK FEED

SUPERIOR FOOD SOURCE Insect larvae of mealworms, black soldier flies, and houseflies are rich in both the macro and micronutrients needed to feed livestock: they contain high levels of protein and fats and are a good source of minerals and vitamins. This makes them highly nutritious for farmed animals, including poultry and pigs, and even fish that readily feed on insects when given the opportunity. Insect nutrient profiles are comparable to or even superior to traditional livestock feed, such as soybeans and fishmeal, and evidence is building for additional benefits from improved health and immunity, as well as the potential reduction in the use of antibiotics in livestock production.

LOWER ENVIRONMENTAL FOOTPRINT Insects are very efficient at converting organic matter into protein and have a high feed conversion rate, which means that they require less food to produce the same amount of protein as traditional livestock. For example, producing 2.2 lb (1 kg) of edible crickets requires 4.8 lb (2.2 kg) of feed, compared to 55 lb (25 kg) of feed for 2.2 lb (1 kg) of beef. This efficiency results in lower greenhouse gas emissions, reduced land and water usage, and less pressure on the environment. Traditional feed sources for livestock, such as cereals, soybean meal, and fishmeal, have a high environmental footprint. Insects can be reared on waste from food production and agricultural by-products. This not only reduces waste management issues, but also promotes a circular and more sustainable agricultural system. Insects can be farmed indoors and vertically, which requires less land and can be done all year round.

LOOKING AHEAD

There are a few challenges to overcome to fully integrate insects into our livestock feed systems. Regulatory frameworks and scaling up production are two of the key areas to address. However, growing interest and investment indicate there is a real opportunity for using insects in livestock production and thus reducing the industry's carbon footprint.

BLACK SOLDIER FLY FARMING

1.

Increasingly, insects such as black soldier flies are being reared on insect farms around the world. Large-scale farms can produce thousands of tons of insect larvae each year.

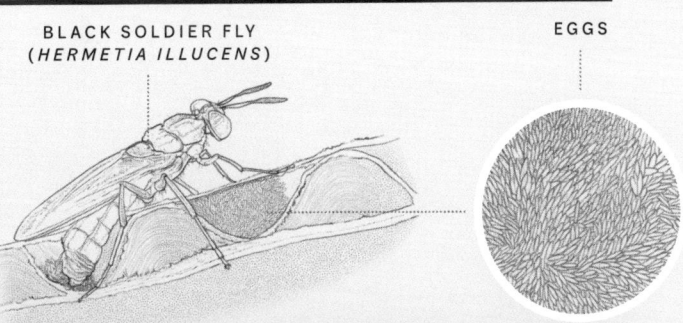

BLACK SOLDIER FLY
(*HERMETIA ILLUCENS*)

EGGS

2.

The life cycle of the black soldier fly lasts five to six weeks. The flies are kept in mating enclosures, where they lay their eggs close to decaying organic matter.

3.

Larvae are able to turn low-value organic products into beneficial proteins. They can be fed on a range of feedstock, including food and animal waste and by-products from agriculture.

4.

The larvae can be harvested live at the fifth instar (stage) and fed to chickens. It is thought that foraging for insects positively impacts the birds' welfare.

5.

Insect larvae can also be separated into protein powder and oils and used to make pellets for animal feed or pet food.

6.

Insect frass (excrement and spent casings) is rich in nutrients and can be used as fertilizer in crop production.

CHICKEN FEEDS
ON LARVAE

*It is often necessary to control insect pests in agriculture.
Chemical insecticides are frequently used, but scientists have
also developed more environmentally sensitive methods, such
as biological control and plant breeding.*

PEST CONTROL IN AGRICULTURE

CHEMICAL CONTROL In the 1940s, DDT seemed to solve the problem of crop pests, but resistant populations of the pests soon appeared—and the same thing happened with the other insecticides that followed DDT. It was also realized that insecticides can contaminate the environment and disrupt biological control.

BIOLOGICAL CONTROL Crop pests can be controlled by working with their natural enemies, such as ladybug predators and parasitoid wasps. These insects can be purchased for mass release in confined spaces such as glasshouses. However, this technique rarely succeeds among field crops because it is difficult to release enough insects to control the pests. Providing natural enemies with food, such as plants with pollen and nectar, can increase their abundance and optimize their impact on insect pests. Noncrop plants can also be planted to host alternative prey, which allows natural enemies to build up in numbers and then switch their attention to pests on the nearby crop.

HOST PLANT RESISTANCE As new crop varieties have been developed, the chemical and physical properties that protect wild plants against pest attack have often been lost. Including wild relatives in plant breeding programs can, however, reestablish plant resistance against pests.

CULTURAL CONTROL Choosing when to sow seeds can affect pest insects. A late-sown crop, for example, might reduce the time available for pest numbers to rise to damaging levels. However, the most effective cultural control is to plant

crop mixtures, because pests may get confused by a mixture of plant volatiles and avoid settling on the crops. Intercropping (growing two or more crops together) is common in the tropics. It reduces pest attacks because the crop mixture mimics natural ecosystems where pest outbreaks are much rarer.

PEST MANAGEMENT This sustainable approach, which delays or avoids the need for insecticide, rests on three simple ideas:

- Establishing "economic thresholds" so that insecticides are only used when yield losses exceed control costs.
- Combining partially effective control measures (biological, cultural, and plant resistance) may often give a greater overall control impact than the sum of the parts.
- Using insecticides carefully to increase the ratio of natural enemies to pests. Attention should be paid to the dosage, choice of insecticide, and timing of the application, for example.

Buckwheat (Polygonaceae)
Flowering buckwheat is planted in a vineyard to encourage the natural enemies of vine pests.

Japanese rhinoceros beetle
(*Allomyrina dichotoma*)
This beetle's characteristic long, forked
horn takes the shape of a letter Y. Males
use it during mating season to defend their
territory. They can lift potential competitors
off the ground and toss them aside.

ORIGINS OF THE *POKÉMON* GAME

As a young boy growing
up in Japan, video game
designer Satoshi Tajiri was
surrounded by a culture of
catching and keeping insects
as pets. He was so fascinated
by the insect world that he
contemplated a career as an
entomologist. Instead, his love
of bugs ultimately inspired
him to create the global
phenomenon known as
Pokémon. Eleven orders
of insects are represented
in the game.

In recent years, there has been a growing interest in keeping insects as pets, particularly among hobby entomologists and those interested in wildlife conservation.

INSECTS AS PETS

Insects are relatively low-maintenance pets, in that they require very little space and minimal care. Looking after them offers a valuable source of education for children, teaching them about the natural world and the biology and behavior of insects in a hands-on manner. A wide variety of insects can be kept as pets.

STICK INSECTS Among the most popular insect pets are stick insects, which come in a range of shapes, sizes, and colors. They are best known for their complex camouflage, which allows them to blend in perfectly with their surroundings. This phenomenon, known as background matching or protective resemblance, can be easily observed in species such as the giant Malaysian leaf insect (*Phyllium giganteum*). Its leaf-shaped body is covered in a pattern of veins, similar to those found in plants. When in motion, it even sways gently, like a leaf moving in the breeze.

COCKROACHES Often portrayed as repulsive and unhygienic in popular media, cockroaches actually make excellent pets. They are low maintenance, social, easy to handle, and surprisingly clean. Madagascar hissing cockroaches (*Gromphadorhina portentosa*) are one of the largest species, and they are popular to keep as a pet because they are wingless and cannot fly. They are, however, exceptional climbers.

BEETLES In Japan, beetles have become cherished pets, particularly among young children. This is partly due to the fact that most apartments do not allow for more traditional pets, such as dogs or cats. Large horned beetles, such as stag beetles (known as *kuwagata* in Japanese) and rhinoceros beetles (*kabutomushi*), reign as the most popular choices.

Insect hotel
The wooden panel provides a habitat to
encourage beneficial insects into the garden.
This illustration is inspired by Tom Massey's
design for the Royal Entomological Society
Garden at RHS Chelsea Flower Show 2023.

Working in domestic gardens and local green spaces provides an ideal opportunity to experience firsthand how biodiversity and insects are being affected by climate change and human disruption.

GARDENS AND GREEN SPACES

FINDING A BALANCE Chemicals are often used to wage indiscriminate warfare on the insects in our green spaces, but instead we can entice beneficial insects, such as pollinators and natural predators, into the garden to control pests. We can also resist the urge to clean up; piles of leaves, decaying plant matter, and dead wood will provide food and habitats for a bustling microcosm of creatures. Plants that are commonly considered weeds—such as dandelions, clover, vetch, and knapweed—are vital food sources for insects, and reducing the frequency of mowing and weeding helps maintain them. We should also embrace the less conventionally charming insects, such as ants and flies, because they do essential work in breaking down organic material and are a valuable food source for other beneficial garden animals, such as frogs and birds.

THOUGHTFUL DESIGN An eclectically designed landscape provides a rich diversity of habitats for insects: think bark mulches, hoggin pathways (gravel, sand, and clay), and log piles. Choosing porous natural landscaping materials, such as dry-stone walling and woven willow fencing, will provide crevices and gaps where insects can shelter. Including a water feature will attract even more wildlife to the garden. Something as simple as leaving depressions in the landscape or having small containers where rainwater can collect provides butterflies, moths, and other insects with somewhere they can hydrate or make their home. Planting perennials and shrubs that flower at different times of the year will extend the seasons when insects can feed and thrive, and selecting a wide range of plants and trees will support a diverse range of insects.

*Before the advent of photography, artistic depictions of
insects that were at the time unknown to science made
important contributions to species identification
and to our knowledge of insect life cycles and behaviors.*

ILLUSTRATING INSECTS FOR SCIENCE

NEW TO SCIENCE From 1699 to 1701, botanical artist Maria Sibylla Merian traveled in the Dutch colony of Surinam, where she studied, reared, and drew insects, some of which were previously unknown to science. The result was her illustrated book *Metamorphosis of the Insects of Surinam*, published in 1705. The first edition described 131 species of insects, including 86 species of butterflies and moths. It was the first illustrated book on South American insects and the first to show the whole life cycle of insects together with their host plants.

Merian initially produced paintings on vellum, which were then etched onto copper plates. The engravings were made at an unusually large scale, which allowed for an exceptional level of detail. Carl Linnaeus identified and named more than 100 animal species, previously unknown in Europe, from her illustrations instead of from specimens. His description of the Menelaus blue morpho butterfly (*Morpho menelaus*, originally named *Papilio menelaus*) cites an illustration from Merian's famous work.

NAMING SPECIES Many insects were named in Merian's honor. These include a subspecies of split-banded owlet butterfly (*Opsiphanes cassina merianae*); a subspecies of the postman butterfly (*Heliconius melpomene meriana*); and a rare butterfly from Panama, *Catasticta sibyllae*. Other insects bearing her name include the Cuban sphinx moth (*Erinnyis merianae*), a genus of mantis known as *Sibylla*, a true bug (*Plisthenes merianae*), and the orchid bee (*Eulaema meriana*).

"The Pineapple Fruit" Plate 2 from *Metamorphosis of the Insects of Surinam* shows the life cycle of the dido longwing butterfly (*Philaethria dido*). The depiction of the butterfly is accurate, but the artist incorrectly shows it feeding on a pineapple in this hand-colored engraving.

Eishōsai Chōki
Woman and Child Catching Fireflies (c.1793)
In this woodblock print, the artist depicts a child's joy at seeing fireflies during an evening stroll.

Vincent Van Gogh
Long Grass with Butterflies (1890)
This painting is from a series titled *Butterflies*, in which the artist depicts butterflies and moths as symbols of escape or freedom.

The earliest insect art is a cave painting from 4000 BCE, near Valencia, Spain, that depicts collecting honey from wild bees. Since then, insects have been used in art symbolically, metaphorically, and decoratively.

INSECTS IN ART

SYMBOLS In medieval Europe, insects appeared as decoration in the margins of manuscripts (mostly of a religious nature), with butterflies often symbolizing the release of the soul. Insects have also been used to represent decay and the transience of worldly things, from the Dutch still-life paintings of the 16th to 18th centuries right up to Damian Hirst's *A Thousand Years* (1990), which features a cow's head on which flies are breeding.

DETAILED PORTRAYALS Intricate studies of insects began in 1665, when British scientist Robert Hooke became the first to generate drawings of fleas, lice, and flies using a microscope. Today, photographer Giles Revell and artist Mark Fairnington continue this work, producing large, detailed images of insects. The work of Turkish artist Ergin Inan includes realistic paintings of insects set in dreamlike backgrounds of ancient Arabic scripts and surreal portraits. Less detailed are the paintings of David Measures and John Walters, who create images in the field, showing the behavior and movement of the insect subject. After the Chernobyl disaster in 1986, scientific illustrator Cornelia Hesse-Honegger painted the asymmetric deformities in shield bugs and other true bugs.

BEAUTY As objects of beauty, insects often appeared in Art Nouveau design, popular between 1890 and 1910. Since the 1980s, French artist Hubert Duprat has created tubes of precious metals and gemstones by offering them to caddisfly larvae, which normally build tubes of sand and other materials (pp.58–59). More recently, Belgian artist Jan Fabre has used the iridescent green wing cases of jewel beetles to create mosaics, high-fashion garments, and the decorative elements of the ceiling at the Royal Palace in Brussels.

Insects make few appearances in literature—as characters or literary devices—until the late 1800s, when popular science began to bring them to the attention of the wider population.

INSECTS IN LITERATURE

INSECTS, REAL AND IMAGINED Prior to the 20th century, insects generally appeared in literature as an aside or in colloquial language, but there are a few notable exceptions. For example, in *Aesop's Fables* (late to mid-6th century BCE), insect characters appear in morality tales such as *The Ant and the Grasshopper*, which advocates the benefits of hard work and planning ahead. In Lewis Carroll's *Alice's Adventures in Wonderland* (1865), a disoriented Alice encounters a hookah-smoking caterpillar who, despite his rudeness, advises her on how to change size to fit in.

MODERN LITERATURE In contemporary novels, the perceived nature of a particular insect is often used in the book's title to indicate a plot or theme. The term "moth," for example, appears in many book titles to signal that the story deals with an unstoppable attraction or extreme fragility. Bees sometimes feature as a focal point in a book's narrative, commonly associated with hard work, the power of communities, and hope, as in *The Secret Life of Bees* (2001) by Sue Monk Kidd. Set within a beehive and based on bee biology, Laline Paull's *The Bees* (2014) is a sinister tale of rebellion in a rigid society in which the main character, Flora, is a worker bee born in the lowest caste.

Fusing reality and fantasy, *The Metamorphosis* (1915) by Franz Kafka is a modern classic in which a traveling salesman wakes up to discover he has become a large insect (which one has been hotly debated). The story reveals how he and his family deal with this unusual transformation. This idea is parodied in Ian McEwan's humorous tale *The Cockroach* (2019), in which a cockroach wakes up to find he is the prime minister of the UK.

ENTOMOLOGISTS IN NOVELS Novels featuring entomologists as protagonists often present the characters as passionate explorers in search of an entomological prize. In *Boxer Beetle* (2010) by Ned Beauman, a cast of dubious characters follow references in a letter from Hitler to hunt for a beetle whose wing marking forms a swastika, while Rachel Joyce's *Miss Benson's Beetle* (2020) tells of a search for a golden beetle that Miss Benson was shown by her father when she was a girl. M. G. Leonard's *Beetle Boy* (2016) is another odyssey, in which Darkus, a young boy, is looking for his entomologist father who has vanished. This leads him to discover a population of highly evolved and intelligent beetles. In E. O. Wilson's semiautobiographical *Anthill* (2010), a young entomologist fights to save the local wilderness from development. His detailed studies of local ant species present the insects' battle for dominance as a metaphor for his own struggle with the developers. This emerging genre of entomological novels will continue to grow as society becomes more aware of the vital role that insects play in maintaining life on Earth.

Fontaine's fable
This illustration is from a c.1910 version of *A Hundred Fables of La Fontaine* by 17th-century author Jean de La Fontaine. Originally one of *Aesop's Fables*, the story reveals how kindness and good deeds are reciprocated.

Entomologists know that insects are incredible,
but many other people have yet to discover this.
Here are a few ways to engage new audiences.

ENGAGING WITH INSECTS

FACTS VS. STORIES Sharing impressive facts—there are more than 1 million insect species recorded in the world, for example—can pique someone's interest. However, telling unusual stories may leave a longer-lasting impression: when an aphid is born, it already has its children growing in the abdomen, so the mother's life conditions directly affect its children and future generations. Eye-catching images, such as a magnified depiction of aphid young inside their mother's abdomen, can help enrich the story. Mainstream media or social media posts can also draw in new audiences.

ACTIVITIES WITH INSECTS Live insects are a great way to trigger conversations with people of any age, whatever their level of knowledge. These could be on display in a terrarium or held in the hand (if appropriate) with experienced supervision. Preserved specimens can also help demonstrate the amazing diversity of insects, including those most closely experienced by humans, such as bed bugs and mosquitoes. Art activities are particularly good for engaging children and young people. They can be used to generate a whole range of insect interpretations, and directed activities such as building a model insect from cardboard tubes (body) and paper straws (legs/antenna) are a great way to share knowledge of insect anatomy.

LASTING MEMORIES The aim with engagement is always that someone will take away a lasting memory. They may leave with a fascinating story to share with someone else, or they may have changed their mind about how to react to the insects they encounter.

**Meadow grasshopper
(*Pseudochorthippus parallelus*)**
Some grasshoppers have a genetic
mutation known as erythrism, which
gives the insect a pinkish-red color.

GLOSSARY

aestivation Dormancy in an animal that takes place in the summer or hot, dry season instead of the winter. Similarly to hibernation, it is characterized by inactivity and lowered metabolic rates.

alate Having wings instead of apterous (wingless).

alien species A species that has been transported outside of its natural range by human activity. (Also termed non-native species.)

arthropod An invertebrate animal that has a segmented body and jointed limbs and that sheds its exoskeleton at intervals. Includes insects, but also crustaceans and arachnids, for example.

biodiversity The variety of all living species and their interactions.

biological control A method of controlling pest species that uses their natural enemies, such as predators and parasitoids, to suppress them.

brood A group of young or juvenile offspring cared for at one time.

cellulose An important structural component in plant cell walls.

clypeus The hardened plate on the front of an insect's head, below the frons and above the labrum.

community science The involvement of volunteers in environmental recording and monitoring processes, for example. (Also termed "citizen science" or "participatory monitoring.")

cuticle The outermost covering of an insect, made primarily of chitin, but also wax and proteins, secreted by epidermal cells.

decomposer A set of organisms that decompose (or break down) dead or decaying organic material. Also used to refer to species that break down

the material externally with chemicals and enzymes.

detrivore A type of decomposer. These species feed on dead and decaying organic matter and detritus, extracting nutrition through consuming and digesting the material

diapause Period of suspended development, growth, or reproduction.

diurnal Active or occurring during the day.

domatia Structures produced by plants to provide a shelter or suitable microenvironment for insects and other arthropods that live in symbiosis with the plant.

ecosystem A biological community or network of interacting species and their physical environment as part of a complex system.

ectotherm An animal that cannot regulate its own body temperature and is dependent on outside sources of heat.

elytra The toughened forewings of some species of insects, such as beetles and earwigs.

entomology The branch of science focused on the study of insects.

exoskeleton The hardened external covering that provides rigidity and structure and protects the internal organs. In insects, it is made of chitin.

flagellum A microscopic thread- or whiplike structure that is involved in propelling or moving a cell.

frons The hardened plate in the middle of the front of an insect's head, between the eyes and above the clypeus.

galea One of the mouthparts of some insects; a flaplike lobe on the outside of the top of the maxillae.

gall An abnormal plant growth caused by an insect or organism.

ganglia Groups of nerve-cell bodies that make up part of the nervous system and nerve signals.

generalist A species that makes use of a wide range of food, habitats, and other environmental resources.

gongylidia Hyphal swellings of fungal colonies cultivated by fungus-growing ant species.

guano The excrement, usually accumulated, of birds and bats. It can make a useful fertilizer.

gyne Reproductive female of social insect species.

hemolymph Blood equivalent in insects and most invertebrates.

hospicide A disinfectant.

instar The developmental stage between successive molts until adulthood and sexual maturity is reached.

invasive alien species A subset of alien or non-native species that establishes (self-sustaining populations) and spreads with adverse impacts on native biodiversity; ecosystems; species; and, in some cases, people. (Also termed invasive non-native species.)

invertebrate A broad umbrella term that includes animals without a vertebral column (or backbone), such as insects and other arthropods.

labrum The hardened plate that forms the "upper lip" of an insect's mouthparts, under the clypeus.

lacinia One of the mouthparts of some insects; a flaplike lobe on the inside of the top of the maxillae.

larva/e Immature stages of insects with complete metamorphosis, such as flies and beetles.

luciferase An enzyme that speeds up the oxidation of luciferin, the light-producing compound that

allows insects such as glow worms to produce bioluminescence.

mandibles Paired mouthparts of some insects and arthropods, equivalent to the jaws. They are used for cutting, biting, and gripping and move at right angles to the body.

maxillae Paired and more mobile mouthparts of some insects and arthropods, set behind the mandibles. In some species, they are like pincers and can be used to manipulate food. In others, they have evolved to form needle- or strawlike mouthparts that allow insects to pierce tissues and suck up liquids.

microorganism A tiny organism of microscopic size. It can exist either as a single-celled individual or as colonies of cells. Examples include bacteria, protozoa, and fungi.

molt/ing The process of shedding the exoskeleton between stages of development in invertebrates.

mutualism Two different species of organism living in a long-term biological association from which both species benefit.

nocturnal Active or occurring during the night.

nymph Immature stages of insects with incomplete metamorphosis, such as dragonflies, damselflies, and grasshoppers.

oviposit To lay an egg or eggs.

pedicel The structure connecting the thorax and the abdomen of an insect.

pinnae External coverings or extensions of the ear in insects and mammals, which typically function to direct sound to the eardrums.

predation Attack and consumption of one animal by another for sustenance.

prepupa The immature inactive and nonfeeding stage of an insect between mature larva and pupa.

pronotum The hardened plate structure that covers all or most of the thorax in some insect species.

provisioning To provide a supply, or cache, of resources for later or gradual access.

pupa/e Immature stage of an insect, between larva and adult.

range Habitat or geographical range where a species can live.

resilin A protein found in the cuticles of insects and other arthropods, particularly in the hinges and wings.

rostrum The projection on the front of some insect heads.

scape The bottom segment of the antennae of some insects.

seta/e Stiff hair- or bristlelike structures.

siphunculi A defining feature of aphids. A pair of these usually tubular, upright, and backward-facing organs is located on the back of the abdomen of aphids. In some species, they may just be a pore.

specialist A species that makes use of a narrow range of food, habitats, and other environmental resources.

spiracle An external respiratory opening, present as pores on an insect's body.

stridulation The rubbing together of body parts, such as legs or wings, to produce a sound.

symbiotic Two different species of organism living in a close and long-term biological association with each other. The relationship can be parasitic, mutualistic, or commensalistic.

taxonomy The branch of science focused on the classification, naming, and description of species.

tentoria The internal support framework of an insect's head.

thermoregulation Ability or action by a species to maintain body temperature with a specific range, even when outside temperatures are outside of this range.

tonotopy The organization of sounds into different receptor locations in the ear or nervous system, based on their frequency (or pitch).

trachea Part of a network of tubules that functions as an insect's respiratory system.

trilobite A group of extinct arthropods known from fossils. They lived in Paleozoic seas and had segmented bodies divided lengthwise along the back into three parts.

troglobite A species that is strictly bound to live in an underground habitat, such as a cave.

type A taxonomic term. A "type" is the particular specimen with which the scientific name of a species is formally associated.

vertex The hardened plates on the top of an insect's head.

viviparity Giving birth to live young instead of laying eggs.

volatiles Mixtures of compounds that can readily evaporate or vaporize under normal conditions.

BIBLIOGRAPHY

pp.14–15 The Insect Body Plan Gullan, P.J., and Cranston, P.S. (2005) *The insects an outline of entomology*. 3rd ed. Oxford: Blackwell Publishing.

Hillyer, J.F., and Pass, G. (2020) "The insect circulatory system: Structure, function, and evolution," *Annual Review of Entomology*, 65(1), pp. 121–143. doi:10.1146/annurev-ento-011019-025003.

pp.22–23 Insect Hearing Bennet-Clark, H.C. (1971) "Acoustics of insect song," *Nature*, 234(5327), pp. 255–259. doi:10.1038/234255a0.

Bailey, W.J. (1993) "The tettigoniid (Orthoptera : Tettigoniidae) ear: Multiple functions and structural diversity," *International Journal of Insect Morphology and Embryology*, 22(2–4), pp. 185–205. doi:10.1016/0020-7322(93)90009-p.

Montealegre-Z., F. *et al.* (2012) "Convergent evolution between insect and mammalian audition," *Science*, 338(6109), pp. 968–971. doi:10.1126/science.1225271.

Woodrow, C., and Montealegre-Z, F. (2023) "Auditory system biophysics in a new species of false-leaf katydid (Tettigoniidae: Pseudophyllinae) supports a hypothesis of broadband ultrasound reception," *Zoologischer Anzeiger*, 304, pp. 94–104. doi:10.1016/j.jcz.2023.04.002.

Warren, B., and Nowotny, M. (2021) "Bridging the gap between mammal and insect ears—a comparative and evolutionary view of sound-reception," *Frontiers in Ecology and Evolution*, 9. doi:10.3389/fevo.2021.667218.

pp.34–35 Mimicry Gilbert, F. (2005) "The evolution of imperfect mimicry," in *Insect Evolutionary Ecology*. Wallingford: CABI Publishing, pp. 231–288.

Chatelain, P. *et al.* (2023) "Müllerian mimicry among bees and Wasps: A review of current knowledge and future avenues of research," *Biological Reviews*, 98(4), pp. 1310–1328. doi:10.1111/brv.12955.

pp.36–37 Glowing In The Dark *Firefly Species Checklist of the USA and Canada (2022) Firefly Atlas*. Available at: **https://www.fireflyatlas.org/firefly-species/firefly-species-checklist**

pp.42–43 Flight Goldsworthy, G.J., and Wheeler, C.H. (2018) *Insect flight*. Boca Raton, FL: CRC Press.

Woiwood, I., Reynolds, D.R., and Thomas, C.D. (2001) *Insect movement: Mechanisms and consequences: Proceedings of the Royal Entomological Society's 20th symposium*. New York: CABI Pub.

pp.48–49 Social Insects: Ants Schultheiss, P. *et al.* (2022) 'The abundance, biomass, and distribution of ants on Earth,' *Proceedings of the National Academy of Sciences*, 119(40). doi:10.1073/pnas.2201550119.

pp.54–55 Eusocial Bees Wcislo, W., and Fewell, J.H. (2017) "Sociality in Bees," in *Comparative Social Evolution*. Cambridge: Cambridge University Press, pp. 50–83.

Roubik, D.W. (2012) "Ecology and Social Organisation of Bees," *Encyclopedia of Life Sciences*. doi:10.1002/9780470015902.a0023596.

pp.56–57 Social Insect-Built Structures Bochynek, T. *et al.* (2021) "Anatomy of a superorganism—structure and growth dynamics of army ant bivouacs," *arXiv, Cornell University*. Available at: **https://arxiv.org/abs/2110.09017v1**

Mujinya, B.B. *et al.* (2014) "Spatial patterns and morphology of termite (Macrotermes Falciger) mounds in the Upper Katanga, D.R. Congo," *Catena*, 114, pp. 97–106. doi:10.1016/j.catena.2013.10.015.

Joseph, G.S. *et al.* (2015) "Termite Mounds mitigate against 50 years of herbivore-induced reduction of functional diversity of savanna woody plants," *Landscape Ecology*, 30(10), pp. 2161–2174. doi:10.1007/s10980-015-0238-9.

Joseph, G.S. *et al.* (2013) "Escaping the flames: Large termitaria as refugia from fire in Miombo Woodland," *Landscape Ecology*, 28(8), pp. 1505–1516. doi:10.1007/s10980-013-9897-6.

Hood, A.S. *et al.* (2020) "Termite Mounds House a diversity of taxa in oil palm plantations irrespective of understory management," *Biotropica*, 52(2), pp. 345–350. doi:10.1111/btp.12754.

Kalko, E.K., Ueberschaer, K., and Dechmann, D. (2006) "Roost structure, modification, and availability in the white-throated round-eared bat, *Lophostoma Silvicolum* (Phyllostomidae) living in active termite nests," *Biotropica*, 38(3), pp. 398–404. doi:10.1111/j.1744-7429.2006.00142.x.

Valdivia-Hoeflich, T., Vega Rivera, J.H., and Stoner, K.E. (2005) "The citreoline trogon as an ecosystem engineer," *Biotropica*, 37(3), pp. 465–467. doi:10.1111/j.1744-7429.2005.00062.x.

Riley, J., Stimson, A.F., and Winch J.M. (1985) "A review of Squamata ovipositing in ant and termite nests," *Herpetological Review*, 16(2), pp. 38–43.

Kistner, D.H. (1969) "The Biology of Termitophiles," in *Biology of Termites*. 1st ed. New York and London: Academic Press, pp. 525–555.

Cai, C. *et al.* (2017) "Early evolution of specialized termitophily in Cretaceous rove beetles," *Current Biology*, 27(8), pp. 1229–1235. doi:10.1016/j.cub.2017.03.009.

pp.62–63 Insect–Insect Interactions Ambele, C.F. *et al.* (2023) "Managing insect services and disservices in Cocoa Agroforestry Systems," *Agroforestry Systems*, 97(6), pp. 965–984. doi:10.1007/s10457-023-00839-x.

Anderson, C.N., and Grether, G.F. (2009) "Interspecific aggression and character displacement of competitor recognition in hetaerina damselflies," *Proceedings of the Royal Society B: Biological Sciences*, 277(1681), pp. 549–555. doi:10.1098/rspb.2009.1371.

Diby, L. *et al.* (2014) "Cocoa Land Health Surveillance: An evidence-based approach to sustainable management of Cocoa landscapes in the Nawa region, South-West Côte d'Ivoire," *World Agroforestry Centre (ICRAF)*, 193.

Drury, J.P., and Grether, G.F. (2014) "Interspecific aggression, not interspecific mating, drives character displacement in the wing coloration of male rubyspot damselflies (hetaerina)," *Proceedings of the Royal Society B: Biological Sciences*, 281(1796), 20141737. doi:10.1098/rspb.2014.1737.

Kastinger, C., and Weber, A. (2001) "Bee-flies (bombylius spp., Bombyliidae, Diptera) and the pollination of flowers," *Flora*, 196(1), pp. 3–25. doi:10.1016/s0367-2530(17)30015-4.

Nash, D.R. *et al.* (2008) "A mosaic of chemical coevolution in a large blue butterfly," *Science*, 319(5859), pp. 88–90. doi:10.1126/science.1149180.

pp.66–67 Brood Parasitism In Bees Danforth, B.N. *et al.* (2019) *The solitary bees: Biology, evolution, conservation*. Princeton, NJ: Princeton University Press.

pp.72–73 Predatory Beetles Bauer, T. (1981) "Prey capture and structure of the visual space of an insect that hunts by sight on the litter layer (*notiophilus biguttatus* F., Carabidae, Coleoptera)," *Behavioral Ecology and Sociobiology*, 8(2), pp. 91–97. doi:10.1007/bf00300820.

Hintzpeter, U., and Bauer, T. (1986) "The antennal setal trap of the ground beetle *loricera pilicornis*: A specialization for feeding on collembola," *Journal of Zoology*, 208(4), pp. 615–630. doi:10.1111/j.1469-7998.1986.tb01527.x.

Baulechner, D. *et al.* (2020) "Convergent evolution of specialized generalists: Implications for phylogenetic and functional diversity of carabid feeding groups," *Ecology and Evolution*, 10(20), 11100. doi:10.1002/ece3.6746.

Digweed, S.C. (1994) "Detection of mucus-producing prey by Carabus nemoralis Mueller and Scaphinotus marginatus Fischer (Coleoptera: Carabidae)." *The Coleopterists' Bulletin*, 48(4), pp. 361–369.

Konuma, J., Nagata, N., and Sota, T. (2010) "Factors determining the direction of ecological specialization in snail-feeding carabid beetles," *Evolution*, 65(2), pp. 408–418. doi:10.1111/j.1558-5646.2010.01150.x.

Kromp, B. (1999) "Carabid beetles in sustainable agriculture: A review on pest control efficacy, cultivation impacts and enhancement," *Agriculture, Ecosystems & Environment*, 74(1–3), pp. 187–228. doi:10.1016/s0167-8809(99)00037-7.

pp.74–75 Aphid Hunters Ninkovic, *V. et al.* (2013) "Ladybird footprints induce aphid avoidance behavior," *Biological Control*, 65(1), pp. 63–71. doi:10.1016/j.biocontrol.2012.07.003.

Roy, H.E., and Pell, J.K. (2000) "Interactions between entomopathogenic fungi and other natural enemies: Implications for biological control," *Biocontrol Science and Technology*, 10(6), pp. 737–752. doi:10.1080/09583150020011708.

pp.76–77 Insects As Herbivores Strong, D.R., Lawton, J.H., and Southwood, R. (1984) *Insects on plants: Community patterns and Mechanisms*. Oxford: Blackwell Scientific Publications.

Bagchi, R. *et al.* (2014) "Pathogens and insect herbivores drive rainforest plant diversity and composition," *Nature*, 506(7486), pp. 85–88. doi:10.1038/nature12911.

pp.84–85 Generalists Morgan, T. *et al.* (2016) "Floral Sonication is an innate behaviour in bumblebees that can be fine-tuned with experience in manipulating flowers," *Journal of Insect Behavior*, 29(2), pp. 233–241. doi:10.1007/s10905-016-9553-5.

Alem, S. *et al.* (2016) "Associative mechanisms allow for social learning and cultural transmission of string pulling in an insect," *PLOS Biology*, 14(10). doi:10.1371/journal.pbio.1002564.

Bridges, A.D. *et al.* (2023) "Bumblebees acquire alternative puzzle-box solutions via Social Learning," *PLOS Biology*, 21(3). doi:10.1371/journal.pbio.3002019.

Barbier, Y. *Atlas Hymenoptera*. Available at: **www.atlashymenoptera.net/pagetaxon.aspx?tx_id=3042**

Raiol, R.L. *et al.* (2021) "Specialist bee species are larger and less phylogenetically distinct than generalists in tropical plant–bee interaction networks," *Frontiers in Ecology and Evolution*, 9. doi:10.3389/fevo.2021.699649.

Schmid-Hempel, P. *et al.* (2007) "Invasion success of the bumblebee, *Bombus terrestris*, despite a drastic genetic bottleneck," *Heredity*, 99(4), pp. 414–422. doi:10.1038/sj.hdy.6801017.

Tang, Q. *et al.* (2018) "Global spread of the German cockroach, Blattella Germanica," *Biological Invasions*, 21(3), pp. 693–707. doi:10.1007/s10530-018-1865-2.

Ali, J.G., and Agrawal, A.A. (2012) "Specialist versus generalist insect herbivores and plant defense," *Trends in Plant Science*, 17(5), pp. 293–302. doi:10.1016/j.tplants.2012.02.006.

Tosh, C.R., Powell, G., and Hardie, J. (2003) "Decision making by generalist and specialist aphids with the same genotype," *Journal of Insect Physiology*, 49(7), pp. 659–669. doi:10.1016/s0022-1910(03)00066-0.

Fericean, L.M. *et al.* (2012) "The behaviour, life cycle and biometrical measurements of Aphis fabae." *Research Journal of Agricultural Science*, 44(4), pp. 31-37.

pp.90–91 Animal Parasites Reeves, W.K., and Lloyd, J.E. (2019) "Louse flies, Keds, and bat flies (hippoboscoidea)," *Medical and Veterinary Entomology*, pp. 421–438. doi:10.1016/b978-0-12-814043-7.00020-0.

Small, R.W. (2005) "A review of Melophagus Ovinus (L.), the Sheep Ked," *Veterinary Parasitology*, 130(1–2), pp. 141–155. doi:10.1016/j.vetpar.2005.03.005.

Barton, S., Virgo, J., and Krenn, H.W. (2023) "The mouthparts of female blood-feeding frog-biting midges (Corethrellidae, Diptera)," *Insects*, 14(5), p. 461. doi:10.3390/insects14050461.

Ambrozio-Assis, A. *et al.* (2018) "Preferences for anuran calls in hematophagous corethrellids (Diptera: Corethrellidae) from Southern Brazil," *Austral Entomology*, 58(3), pp. 622–628. doi:10.1111/aen.12376.

Onmaz, A.C. *et al.* (2012) "Vectors and vector-borne diseases of horses," *Veterinary Research Communications*, 37(1), pp. 65–81. doi:10.1007/s11259-012-9537-7.

Ashworth, M., Falk, S., and Clements, D.K. (2023) "The genome sequence of the thick-headed fly, *myopa tessellatipennis* (motschulsky, 1859)," *Wellcome Open Research*, 8, p. 115. doi:10.12688/wellcomeopenres.19108.1.

pp.98–99 Insects In The Home https://www.english-heritage.org.uk/learn/conservation/clothes-moth-research/understanding-clothes-moths/

pp.102–103 Habitats Wilson, E.O. (1987) "The little things that run the world* (the importance and conservation of invertebrates)," *Conservation Biology*, 1(4), pp. 344–346. doi:10.1111/j.1523-1739.1987.tb00055.x.

Hance, J. (2011) "The value of the little guy, an interview with Tyler Prize-winning entomologist May Berenbaum," *Mongabay*, April 6. Available at: **https://news.mongabay.com/2011/04/the-value-of-the-little-guy-an-interview-with-tyler-prize-winning-entomologist-may-berenbaum/**

pp.108–109 Tropical Forests Basset, Y. *et al.* (2012) "Arthropod diversity in a tropical forest," *Science*, 338(6113), pp. 1481–1484. doi:10.1126/science.1226727.

Ellwood, M.D., and Foster, W.A. (2004) "Doubling the estimate of invertebrate biomass in a rainforest canopy," *Nature*, 429(6991), pp. 549–551. doi:10.1038/nature02560.

pp.112–113 Soil *FAO Soils Portal: Facts and Figures, Food and Agriculture Organization of the United Nations*. Available at: **https://www.fao.org/soils-portal/soil-biodiversity/facts-and-figures/en/**

Orgiazzi, A. *et al.* (2016) *Global Soil Biodiversity atlas*. Luxembourg: European Commission, Publications Office of the European Union.

No Dig. Charles Dowding. Available at: **www.charlesdowding.co.uk/**

pp.122–123 Ponds And Puddles *Mud-puddling* (2024) Wikipedia. Available at: **https://en.wikipedia.org/wiki/Mud-puddling**

Jones, C. (2024) Restoring "ghost ponds," Farm Wildlife. Available at: **https://farmwildlife.info/2022/03/10/restoring-ghost-ponds/**

pp.124–125 Marshland Lambret, P., Cohez, D., and Janczak, A. (2009) "*Lestes macrostigma* (Eversmann, 1836) en Camargue et en Crau (Département des Bouches-du-Rhône) (Odonata, Zygoptera, Lestidae)," *Martinia*, 25(2), pp. 51–65.

Martinou, A.F., and Roy, H.E. (2018) "From local strategy to global frameworks: effects of invasive non-native species on health and well-being," in *Invasive Species and Human Health*. Wallingford: CABI Publishing, pp. 11–22.

Martinou, A.F. *et al.* (2020) "A call to arms: Setting the framework for a code of Practice for Mosquito Management in European wetlands," *Journal of Applied Ecology*, 57(6), pp. 1012–1019. doi:10.1111/1365-2664.13631.

Encyclopedic Entry: Marsh. National Geographic | Education. Available at: **https://education.nationalgeographic.org/resource/marsh/**

Aedes detritus/Aedes coluzzii—current known distribution (*2023*) European Centre for Disease Prevention and Control and European Food Safety Authority. Available from: **https://ecdc.europa.eu/en/disease-vectors/surveillance-and-disease-data/mosquito-maps**

pp.126–127 Marine Habitats Cheng, L. (1976) *Marine insects*. Amsterdam: North-Holland Pub. Co.

Cheng, L. (1977) "The elusive sea bug Hermatobates (Heteroptera)," *Pan-Pacific Entomologist*, 53, pp. 87–97.

Cheng, L., and Hashimoto, H. (1978) "The marine midge pontomyia (Chironomidae) with a description of females of p.oceana tokunaga," *Systematic Entomology*, 3(3), pp. 189–196. doi:10.1111/j.1365-3113.1978.tb00115.x.

Cheng, L., and Collins, J.D. (1980) "Observations on behavior, emergence and reproduction of the Marine Midges Pontomyia (Diptera: Chironomidae)," *Marine Biology*, 58(1), pp. 1–5. doi:10.1007/bf00386872.

Cheng, L. (2009) "Marine insects," *Encyclopedia of Insects*, pp. 600–604. doi:10.1016/b978-0-12-374144-8.00167-3.

Cheng, L., and Mishra, H. (2022) "Why did only one genus of insects, halobates, take to the high seas?," *PLOS Biology*, 20(4). doi:10.1371/journal.pbio.3001570.

Huang, D., and Cheng, L. (2011) "The flightless marine midge Pontomyia (Diptera: Chironomidae): Ecology, distribution, and Molecular Phylogeny," *Zoological Journal of the Linnean Society*, 162(2), pp. 443–456. doi:10.1111/j.1096-3642.2010.00680.x.

pp.130–131 Urban Environments Baldock, K.C. *et al.* (2015) "Where is the UK's pollinator biodiversity? the importance of urban areas for flower-visiting insects," *Proceedings of the Royal Society B: Biological Sciences*, 282(1803), p. 20142849. doi:10.1098/rspb.2014.2849.

Anderson, M., Rotheray, E.L., and Mathews, F. (2023) "Marvellous moths! pollen deposition rate of bramble (*Rubus futicosus* L. agg.) is Greater at night than day," *PLOS ONE*, 18(3). doi:10.1371/journal.pone.0281810.

Knop, E. *et al.* (2017) "Artificial light at night as a new threat to pollination," *Nature*, 548(7666), pp. 206–209. doi:10.1038/nature23288.

Boyes, D.H. *et al.* (2021) "Street lighting has detrimental impacts on local insect populations," *Science Advances*, 7(35). doi:10.1126/sciadv.abi8322.

Macgregor, C.J. *et al.* (2016) "The Dark Side of street lighting: Impacts on moths and evidence for the disruption of nocturnal pollen transport," *Global Change Biology*, 23(2), pp. 697–707. doi:10.1111/gcb.13371.

Peach Blossom. Butterfly Conservation. Available at: **https://butterfly-conservation.org/moths/peach-blossom**

Red Mason Bee (*2022*) Bumblebee Conservation Trust. Available at: **https://www.bumblebeeconservation.org/redmasonbee/**

Red Mason Bee. The Wildlife Trusts. Available at: **https://www.wildlifetrusts.org/wildlife-explorer/invertebrates/bees-and-wasps/red-mason-bee**

Theodorou, P. *et al.* (2020) "Urban areas as hotspots for bees and pollination but not a panacea for all insects," *Nature Communications*, 11(1). doi:10.1038/s41467-020-14496-6.

McKinney, M.L. (2008) "Effects of urbanization on species richness: A review of plants and animals," *Urban Ecosystems*, 11(2), pp. 161–176. doi:10.1007/s11252-007-0045-4.

Gardiner, M.M., Burkman, C.E., and Prajzner, S.P. (2013) "The value of urban vacant land to support arthropod biodiversity and Ecosystem Services," *Environmental Entomology*, 42(6), pp. 1123–1136. doi:10.1603/en12275.

Splitt, A. *et al.* (2021) "Keep trees for bees: Pollen collection by *Osmia bicornis* along the urbanization gradient," *Urban Forestry & Urban Greening*, 64, p. 127250. doi:10.1016/j.ufug.2021.127250.

pp.132–133 Islands Bartlett, J.C., Convey, P., and Hayward, S.A. (2020) "Surviving the antarctic winter—life stage cold tolerance and ice entrapment survival in the invasive chironomid midge Eretmoptera Murphyi," *Insects*, 11(3), p. 147. doi:10.3390/insects11030147.

Hayward, S.A. *et al.* (2007) "Slow dehydration promotes desiccation and freeze tolerance in the antarctic midge *belgica antarctica*," *Journal of Experimental Biology*, 210(5), pp. 836–844. doi:10.1242/jeb.02714.

Kelley, J.L. *et al.* (2014) "Compact genome of the antarctic midge is likely an adaptation to an extreme environment," *Nature Communications*, 5(1). doi:10.1038/ncomms5611.

Kozeretska, I. *et al.* (2021) "*belgica antarctica* (Diptera: Chironomidae): A natural model organism for extreme environments," *Insect Science*, 29(1), pp. 2–20. doi:10.1111/1744-7917.12925.

Matthews, T.J., and Triantis, K. (2021) "Island biogeography," *Current Biology*, 31(19), pp. R1201-R1207. doi:10.1016/j.cub.2021.07.033.

Sugiura, S. (2010) "Can Hawaiian carnivorous caterpillars attack invasive ants or vice versa?," *Nature Precedings* [Preprint]. doi:10.1038/npre.2010.5374.1.

Teets, N.M., and Denlinger, D.L. (2014) "Surviving in a frozen desert: Environmental stress physiology of terrestrial Antarctic arthropods," *Journal of Experimental Biology*, 217(1), pp. 84–93. doi:10.1242/jeb.089490.

pp.134–135 Deserts Hamilton, W.J., and Seely, M.K. (1976) "Fog basking by the Namib Desert beetle, *Onymacris unguicularis*," *Nature*, 262(5566), pp. 284–285. doi:10.1038/262284a0.

O'Donnell, M.J. (1977) "Site of water vapor absorption in the desert cockroach, *arenivaga investigata*.," *Proceedings of the National Academy of Sciences*, 74(4), pp. 1757–1760. doi:10.1073/pnas.74.4.1757.

Potter, K., Davidowitz, G., and Woods, H.A. (2009) "Insect eggs protected from high temperatures by limited homeothermy of plant leaves," *Journal of Experimental Biology*, 212(21), pp. 3448–3454. doi:10.1242/jeb.033365.

Smolka, J. *et al.* (2012) "Dung beetles use their dung ball as a mobile thermal refuge," *Current Biology*, 22(20), pp. R863-864. doi:10.1016/j.cub.2012.08.057.

Wehner, R., Marsh, A.C., and Wehner, S. (1992) "Desert ants on a thermal tightrope," *Nature*, 357(6379), pp. 586–587. doi:10.1038/357586a0.

pp.138–139 Caves Culver, D.C., and Pipan, T. (2019) *The biology of caves and other subterranean habitats*. 2nd ed. Oxford: Oxford University Press.

Resh, V.H., Cardé, R.T., and Howarth, F.G. (2009) "Cave Insects," in *Encyclopedia of insects*. Amsterdam: Elsevier/Academic Press, pp. 139–143. doi:10.1016/b978-0-12-374144-8.00047-3

Mammola, S. (2018) "Finding answers in the dark: Caves as models in Ecology Fifty Years after Poulson and white," *Ecography*, 42(7), pp. 1331–1351. doi:10.1111/ecog.03905.

von Byern, J. *et al.* (2016) "Characterization of the fishing lines in Titiwai (=arachnocampa Luminosa Skuse, 1890) from New Zealand and Australia," *PLOS ONE*, 11(12). doi:10.1371/journal.pone.0162687.

p.147 Ephemeroptera Barber-James, H.M. *et al.* (2007) "Global diversity of mayflies (Ephemeroptera, Insecta) in freshwater," *Hydrobiologia*, 595(1), pp. 339–350. doi:10.1007/s10750-007-9028-y.

Brittain, J.E., and Sartori, M. (2003) "Ephemeroptera," in *Encyclopedia of Insects*. Amsterdam: Academic Press, pp. 373–380.

Elliott, J.M., Humpesch, U.H., and Macan, T.T. (2012) *Mayfly larvae (ephemeroptera) of Britain and Ireland Keys and a review of their ecology*. Ambleside, Cumbria: Freshwater Biological Association.

pp.150–151 Orthoptera Greenfield, M. (2012) "Songs of Love, Orthoptera-style," *PLoS Biology*, 10(2). doi:10.1371/journal.pbio.1001275.

Bennet-Clark, H.C., and Bailey, W.J. (2002) "Ticking of the clockwork cricket: The role of the Escapement Mechanism," *Journal of Experimental Biology*, 205(5), pp. 613–625. doi:10.1242/jeb.205.5.613.

Sarria-S, F.A. *et al.* (2014) "Shrinking wings for ultrasonic pitch production: Hyperintense ultra-short-wavelength calls in a new genus of Neotropical katydids (Orthoptera: Tettigoniidae)," *PLoS ONE*, 9(6). doi:10.1371/journal.pone.0098708.

Bennet-Clark, H.C. (1970) "The mechanism and efficiency of sound production in mole crickets," *Journal of Experimental Biology*, 52(3), pp. 619–652. doi:10.1242/jeb.52.3.619.

Davranoglou, L., Taylor, G.K., and Mortimer, B. (2023) "Sexual selection and predation drive the repeated evolution of stridulation in heteroptera and other arthropods," *Biological Reviews*, 98(3), pp. 942–981. doi:10.1111/brv.12938.

pp.154–155 Phasmida Bian, X., Elgar, M.A., and Peters, R.A. (2015) "The swaying behavior of *extatosoma tiaratum*: Motion camouflage in a stick insect?," *Behavioral Ecology*, 27(1), pp. 83–92. doi:10.1093/beheco/arv125.

Brock, P.D., and Büscher, T.H. (2022) *Stick and leaf-insects of the world phasmids*. Verrieres le buisson, France: NAP Editions.

Freitas, S. *et al.* (2023) "Evidence for cryptic sex in parthenogenetic stick insects of the genus timema," *Proceedings of the Royal Society B: Biological Sciences*, 290(2007). doi:10.1098/rspb.2023.0404.

Hennemann, F.H., and Conle, O.V. (2008) "Revision of Oriental phasmatodea: The Tribe pharnaciiniGünther, 1953, including the description of the world's longest insect, and a survey of the family Phasmatidae Gray, 1835 with keys to the subfamilies and tribes (phasmatodea: "anareolatae": Phasmatidae)," *Zootaxa*, 1906(1), pp. 1–316. doi:10.11646/zootaxa.1906.1.1.

Jaron, K.S. *et al.* (2022) "Convergent consequences of parthenogenesis on stick insect genomes," *Science Advances*, 8(8). doi:10.1126/sciadv.abg3842.

Morgan-Richards, M., Langton-Myers, S.S., and Trewick, S.A.

(2019) "Loss and gain of sexual reproduction in the same stick insect," *Molecular Ecology*, 28(17), pp. 3929–3941. doi:10.1111/mec.15203.

Vickery, V.R. (1993) "Revision of *timema scudder* (PHASMATOPTERA: Timematodea) including three new species," *The Canadian Entomologist*, 125(4), pp. 657–692. doi:10.4039/ent125657-4.

pp.164–165 Hymenoptera Campbell, H., and Blanchard, B. (2023) *Ants: A Visual Guide*. Princeton, NJ: Princeton University Press.

van Noort, S., and Broad, G. (2024) *Wasps of the World: A Guide to Every Family*. Princeton, NJ: Princeton University Press.

Packer, L. (2023) *Bees of the World: A Guide to Every Family*. Princeton, NJ: Princeton University Press.

Sumner, S. (2022) *Endless Forms: Why We Should Love Wasps*. HarperCollins UK.

p.172 Siphonaptera Durden, L.A., and Hinkle, N.C. (2019) 'Fleas (Siphonaptera),' in *Medical and veterinary entomology*. Cambridge, MA: Academic Press, pp. 145–169.

Krasnov, B.R. (2008) *Functional and evolutionary ecology of Fleas: A model for ecological parasitology*. Cambridge: Cambridge University Press. doi:10.1017/cbo9780511542688.

Griffiths, H.J. (1978) *A Handbook of Veterinary Parasitology: Domestic Animals of North America*. Minneapolis: University of Minnesota Press.

pp.180–181 What Do We Observe And Why? Harrington, R. *et al.* (2012) "The Rothamsted Insect Survey: old traps, new tricks," *Aspects of Applied Biology*, 117, pp. 157–164.

pp.182–183 Community Science State of nature 2023—report on the UK's current biodiversity (2023) *State of Nature*. Available at: **https://stateofnature.org.uk/**

van Klink, R. *et al.* (2022) "Emerging technologies revolutionise insect ecology and monitoring," *Trends in Ecology & Evolution*, 37(10), pp. 872–885. doi:10.1016/j.tree.2022.06.001.

Fit counts: Help us monitor pollinators. PoMS. Available at: **https://ukpoms.org.uk/fit-counts**

iNaturalist. Available at: **https://www.inaturalist.org/**

Powney, G.D. (2019) "Widespread losses among pollinating insects in Britain," *UK Centre for Ecology & Hydrology*, March 26. Available at: **https://www.ceh.ac.uk/news-and-media/news/widespread-losses-among-pollinating-insects-britain**

Powney, G.D. *et al.* (2019) "Widespread losses of pollinating insects in Britain," *Nature Communications*, 10(1). doi:10.1038/s41467-019-08974-9.

pp.184–185 Taxonomy And Systematics Erwin, T.L. (1983) "Tropical Forest Canopies: The last biotic frontier," *Bulletin of the Entomological Society of America*, 29(1), pp. 14–20. doi:10.1093/besa/29.1.14.

International Commission on Zoological Nomenclature (ICZN). Available at: **https://www.iczn.org/**

Linnæus, C. (1758) *Systema naturæ per regna tria naturæ, secundum classes, ordines, genera, species, cum characteribus, differentiis, synonymis, locis Vol 1*. Stockholm: Holmiae, Impensis Direct, Laurentii Salvii.

Tihelka, E. *et al.* (2021) "The evolution of Insect Biodiversity," *Current Biology*, 31(19), pp. R1299–R1311. doi:10.1016/j.cub.2021.08.057.

Ashworth, J. (2023) "Insect trapped in amber reveals the evolutionary battles of ancient Europe," *Natural History Museum*, November 13. Available at: **https://www.nhm.ac.uk/discover/news/2023/november/insect-trapped-amber-reveals-evolutionary-battles-ancient-europe.html**

pp.194–195 Insect Decline Lawton, J. et al. (2010) *Making Space for Nature: a review of England's wildlife sites and ecological network*. Report to the Department for Environment, Food and Rural Affairs (Defra). Available at: **https://webarchive.nationalarchives.gov.uk/ukgwa/20130402170324mp_/http://archive.defra.gov.uk/environment/biodiversity/documents/201009space-for-nature.pdf**

The state of the UK's butterflies: 2022 report (2022) *Butterfly Conservation*. Available at: **https://butterfly-conservation.org/sites/default/files/2023-01/State%of%UK%Butterflies%2022%Report.pdf**

pp.198–199 Alien Insects Roy, H. E. *et al.* (2024) "IPBES Invasive Alien Species Assessment: Summary for Policymakers." *Zenodo*. doi: 10.5281/zenodo.10521002.

pp.202–203 Managing Urban Habitats Brownbill, A., and Dutton, A. (2019) "UK natural capital: urban accounts." Office of National Statistics (ONS). Available at: **https://www.ons.gov.uk/economy/environmentalaccounts/bulletins/uknaturalcapital/urbanaccounts**

Salisbury, A. *et al.* (2015) "Editor's choice· Enhancing gardens as habitats for flower-visiting aerial insects (pollinators): Should we plant native or exotic species?," *Journal of Applied Ecology*, 52(5), pp. 1156–1164. doi:10.1111/1365-2664.12499.

Boyes, D.H. *et al.* (2020) "Is light pollution driving moth population declines? A review of causal mechanisms across the life cycle," *Insect Conservation and Diversity*, 14(2), pp. 167–187. doi:10.1111/icad.12447.

pp.204–205 Restoring Natural Habitats Young, M.R., and Barbour, D.A. (2004) "Conserving the new forest burnet moth (*Zygaena viciae* ([Denis and schiffermueller])) in Scotland; responses to grazing reduction and consequent vegetation changes," *Journal of Insect Conservation*, 8(2–3), pp. 137–148. doi:10.1023/b:jico.0000045811.28261.d1.

The Great Fen. Available at: **https://www.greatfen.org.uk/**

The Wildlife Trust for Lancashire, Manchester and North Merseyside (2023) "Rare manchester argus butterflies flourishing after reintroduction," *The Wildlife Trust*, June 15. Available at: **https://www.lancswt.org.uk/news/rare-manchester-argus-butterflies-flourishing-after-reintroduction**

pp.228–229 In Art Carlson, O. (2019) "Microcosms: An examination of insects in 17th-century Dutch still lifes," *Scholarly Horizons: University of Minnesota, Morris Undergraduate Journal*, 6(2). doi:10.61366/2576-2176.1079.

pp.232–233 Engaging With Insects *How I learned to Stop Worrying and Love the Bug* (2019) Outside/In Radio, October 24. Available at: **http://outsideinradio.org/transcript-how-i-learned-to-stop-worrying-and-love-the-bug**

INDEX

INDEX OF INSECTS

ABOUT THE AUTHORS

ASHLIE AKE is a graduate student in the Veterinary Biomedical Science program at Kansas State University, investigating the effects of microclimate change on insect prevalence and vector competence. Her passion for insects has led her through diverse experiences, from researching bees and sampling mites to rearing beetles and inspiring others by teaching zoology.

LUCY ALFORD is a lecturer in Invertebrate Physiology at the University of Bristol, with interests in ecophysiology and conservation ecology. Her research focuses on the interplay between landscape and climate, how this impacts insects, and how this can inform landscape management and restoration to encourage and conserve beneficial insects.

GUDBJORG INGA (GIA) ARADOTTIR is an entomologist and has been the Treasurer and Trustee of the Royal Entomological Society since 2020. Formerly a research scientist specializing in insect pests in agriculture, she now focuses on innovation and is a consultant and advisor to start-ups, research organizations, and government departments.

SARAH ARNOLD is an applied entomology researcher at NIAB, a UK horticultural research organization. She studies the behavior and ecology of horticulturally important insects, such as pests, natural enemies, and pollinators, and their interactions with the environment.

GAIL ASHTON is a nature photographer, author, and illustrator specializing in insects. She talks incessantly about insects, writes books and articles about them, and is a contributing photographer for the RSPB. Gail promotes the conservation of the UK's rare insect species and their habitats. She finds it impossible to choose a favorite insect, but the bloody-nosed beetle is high on the list.

DIMITRIOS N. AVTZIS is Research Director in forest entomology at the Forest Research Institute, Hellenic Agricultural Organization Demeter, Greece. After two decades, tens of manuscripts, and countless experiences and unforgettable moments, he still has the same (or more) passion and enthusiasm for every aspect of the six-legged wonders of evolution as he had in Vienna back in 2003.

JENNIFER BANFIELD-ZANIN is editorial coordinator at the Royal Entomological Society, where she supports the society's books, journals, and membership magazine portfolios. An entomologist with a background in applied research, she obtained her PhD from Imperial College London and has research interests in insect–plant interactions and pest population dynamics.

HELEN BARBER-JAMES is Senior Curator of Natural Sciences, National Museums Northern Ireland. Her research focus is on insects inhabiting freshwater ecosystems. She was previously head of the Department of Freshwater Invertebrates at the Albany Museum in Grahamstown, South Africa, before moving to the UK in 2022.

MEGHAN BARRETT is an Assistant Professor of Biology at Indiana University and director of the Insect Welfare Research Society. She has authored more than 30 publications on insect neuroscience, physiology, welfare, and entomological ethics.

MAX BLAKE is Head of Entomology at Forest Research, the research agency of the Forestry Commission. Max leads a team of 30 entomologists and PhD students to provide advice, diagnostics, policy direction, and practical research to DEFRA, the Forestry Commission, and devolved administrations on managing and surveying forest insects.

OCTAVIA BRAYLEY is a PhD researcher at the University of Birmingham and British Antarctic Survey, where she is investigating the ecological effects of an invasive species of insect in Antarctica. Octavia is also a part-time science tutor, STEM ambassador, school governor, and co-head of education and outreach at the UK Polar Network.

JOANNA BREBNER is an arthropod lover and entomologist who has spent most of her career watching bees navigate better than she can. She is a Fyssen Foundation postdoctoral fellow in the BeeAntCe team, University Paul Sabatier, Toulouse III (France), after previously completing a PhD with Professor Chittka at Queen Mary University of London.

HEATHER CAMPBELL is a Senior Lecturer in Entomology at Harper Adams University. She is a National Geographic Explorer, with research expertise in insect biodiversity, ecology, and conservation, specializing in social insects in arid ecosystems. She has written several books, including *Ants: A Visual Guide* and *30-Second Ecology*.

LANNA CHENG is editor of the only reference book on marine insects, published in 1976. She is probably the world expert on *Halobates*, the only insect genus known to have conquered the open ocean, where five species spend all their lives at the sea–air interface far away from land.

ANDREW CHICK is a lecturer at the University of Cumbria, where he teaches entomology to zoology and

forensic science students. He was awarded the Trail-Crisp Award by the Linnean Society of London for his work on the slide mounting media used in entomological micro technique.

CLAIRE CRESSWELL is a lecturer in Wildlife Ecology & Conservation and Agriculture at University Centre Sparsholt. They are also a wildlife surveyor and consultant through their own Cresswell Wildlife Consultancy, regularly monitoring and researching insects. They have an avid interest for their interactions with grassland plant species and their traits.

JORDAN CUFF is a molecular ecological entomologist—someone who uses DNA and other molecules to study the ecology of insects and other invertebrates. He is particularly interested in the interactions between invertebrate predators and their prey, and what drives predators to eat what they eat.

DARRON A. CULLEN is a lecturer in Biological Sciences at the University of Hull, where he teaches zoology, genetics, and animal behavior. His research investigates the molecular and physiological control of behavior, especially in locusts, crickets, and mantids. He is a Fellow of the Royal Entomological Society.

BRÓNAGH D'ARCY COBAIN is an ecologist based in Edinburgh, Scotland. She has a keen interest in entomology, with a focus on behavioral ecology and conservation management. Her passion for invertebrates was

encouraged throughout her childhood while being raised in Glasgow by her grandmother.

ELEANOR DRINKWATER is a lecturer at Writtle University College. She has a background in emergent invertebrate behavior and defense, but her research is currently focused on commercial invertebrate farming methods and human perceptions of the invertebrate food and feed industry.

HELMUT VAN EMDEN is Emeritus Professor of Horticulture at Reading University, where he taught agricultural entomology for 40 years. He has been involved with the subject in 27 countries. Among his 240+ publications are 13 books, including *Handbook of Agricultural Entomology*, (2013).

MICHAEL S. ENGEL is an authority on living and fossil insects, with expertise on the evolution of bees and termites. He is the author of *Innumerable Insects* (2018) and co-author of *Evolution of the Insects* (2005).

MARION ENGLAND is a research fellow in vector ecology at The Pirbright Institute, where she leads research on insects that carry animal and human disease, particularly biting midges. Her work focuses on understanding the ecology and behaviors of these insects and the impacts of climate change on disease spread.

SEELAVARN GANESHAN holds a PhD in Applied Entomology from the University of London. He has over 40 years

of experience in pest management in sugarcane, potato, tomato, maize, and other crops in several countries. He is passionate about integrated pest management, biological control, biodiversity conservation, agroecology, and habitat manipulation.

BEULAH GARNER is an entomologist specializing in the classification and evolution of Coleoptera (beetles). She has worked in remote parts of the tropical world, researching their incredible diversity and to discover new species. She is passionate about promoting and communicating insect science and conserving insects and their habitats.

DION GARRETT is an entomologist at Rothamsted Research, with expertise in insect biology, insect identification, molecular techniques, and integrated pest management. He has always had a fascination for the natural world and strives to conserve it. His favorite insect from childhood is the great diving beetle.

FRANCIS GILBERT has been at Nottingham University throughout his career, working mostly but not exclusively with insects. His special focus is on the evolutionary biology of the hoverflies (Syrphidae).

ALEXIS GKANTIRAGAS is a PhD student at Queen Mary University of London, studying the molecular biology of bumblebees. They have published several peer-reviewed papers, opinion pieces, and flash fiction, and their work has been covered by outlets such as FastCompany and TechXplore.

H. CHARLES J. GODFRAY is Director of the Oxford Martin School, and former Hope Professor of Entomology at Oxford University. He is a population biologist who has often worked on fundamental and applied research questions involving insects. He has a strong professional and natural history interest in parasitoid wasps.

JIM HARDIE is an Emeritus Professor at Imperial College London and an insect ecophysiologist with a fondness for aphids and blow-fly maggots. Jim has been a Fellow of the Royal Entomological Society for more than 40 years and has served as president. He is currently the society's Resident Entomologist.

RICHARD HARRINGTON worked with the Rothamsted Insect Survey to understand factors affecting aphid populations so that predictions of pest problems could be made and control measures only implemented when necessary. He is an Honorary Fellow of the Royal Entomological Society and has served on its Council and many of its committees.

ADAM HART is an entomologist, ecologist, and conservation scientist with a special interest in social insects. He is Professor of Science Communication at Gloucestershire University, where he combines research and teaching with public science communication, including broadcasting and writing for general audiences.

WILL HAWKES is an insect migration scientist from North Wales. He spent his childhood chasing bugs and,

now in his 20s, nothing has changed. His time is spent traveling to the high mountains or isolated islands, exploring the lives of migratory insects and helping share the tales of their remarkable journeys.

SIMON HELLEMANS is a passionate Belgian entomologist. He taught summer schools on plant-insect interactions and carried out sampling missions in tropical ecosystems. Combining classical methods in ecology and next-generation sequencing, his studies are primarily focused on termites' evolutionary history, microbial symbioses, and reproductive strategies.

DIANA AMEYALLI (AME) HERNÁNDEZ MÁRQUEZ is fascinated by insects, but especially passionate about bees. She has researched them in Mexico and Europe, focusing on the effects of climate change on bee behavior during her MSc in Evolutionary & Behavioural Ecology. She hopes to keep learning about bees for many years.

THOMAS HESSELBERG is a lecturer at the University of Oxford. His research expertise is on the behavioral ecology of insects and spiders, with a focus on behavioral adaptations in orb-web spiders to a range of habitats, including caves, islands, and urban areas. He is a Fellow of the Royal Entomological Society.

STEPHEN HIGGS has conducted research on mosquito–virus–vertebrate interactions since 1985, with over 200 publications. He received the Hoogstraal

Medal for outstanding achievement in medical entomology and was a co-recipient of the Bial Award for most significant publication over a 10-year period. He is director of the Biosecurity Research Institute.

WERNER E. HOLZINGER lives in Graz, Austria, and has a lifelong interest in observing nature and animals, especially insects. His main fields of research are dragonflies, planthoppers, and leafhoppers. He is the head of the Oekoteam, an environmental consultancy, and a Professor at the University of Graz, teaching nature conservation.

AMELIA HOOD is an agricultural ecologist and social scientist. She conducted her PhD in oil palm plantations in Indonesia, where her fascination with social insects, and termites in particular, was sparked. She is currently working as a postdoctoral researcher at the University of Reading.

DAEGAN INWARD is a research entomologist at Forest Research (Forestry Commission), investigating the ecology and impacts of invasive forest pests, and their interactions with UK host trees in particular. He started his entomological journey at the Natural History Museum London, studying the biogeography and evolution of dung beetles and termites.

HAYLEY JONES is Principal Entomologist at the Royal Horticultural Society. Her work focuses on giving advice to gardeners on managing plant health problems in a sustainable

and wildlife-friendly way and encouraging them to love and enjoy the invertebrate life in their gardens. Hayley's favorite insects are moths.

KELLY JOWETT is an entomologist at Rothamsted Research. She is working with farmers to apply farm management that will conserve beneficial beetles, supporting sustainable agriculture. She has a background in conservation, ecology, forestry, and food security. She has had a special interest in beetles from childhood but loves all invertebrates.

TIFFANY KI is a conservation scientist interested in understanding biodiversity and ecosystem responses to anthropogenic pressures and relating these dynamics to consider how best to reconcile continued human development and conservation. She is particularly fascinated by insects, museum collections, and their use to understand biodiversity change and tackle conservation challenges.

MERI LÄHTEENARO has been fascinated by insects since childhood. Currently pursuing a PhD at the Swedish Museum of Natural History, she is exploring the diversity and systematics of the twisted-winged parasite genus *Stylops*. Previously, Meri has contributed to the regional red list assessment of the order Strepsiptera in Finland.

PAUL LATHAM is a retired Salvation Army officer who worked in Africa for more than 20 years, mostly in farmer training. More recently, he has worked

with the Salvation Army on a bee-keeping and edible caterpillar project in Kongo Central province, DRC. He is married and now lives in Scotland.

OWEN LEWIS is Professor of Ecology at the Department of Biology, University of Oxford. He studies insect biodiversity in a range of ecosystems, especially tropical rainforests, with a focus on the response of insect communities to environmental change and the consequences for species interactions and ecosystem services.

SOPHIE MALLET is a Cardiff University PhD student whose work uses innovative methods to understand the thermal ecology of ants across continents. Sophie earned a BSc in Biological Sciences from Imperial College London in 2021 and researched the nutritional mutualism between fungus and leafcutter ants with the University of Copenhagen.

ANGELIKI F. (KELLY) MARTINOU is the Head Entomologist at the Joint Services Health Unit, British Forces Cyprus. She is inspired by insects and tries to understand how anthropogenic pressures impact insects of medical importance, such as mosquitoes, but also beneficial species.

TOM MASSEY designs award-winning gardens for private and commercial clients, as well as for festivals and shows in the UK and overseas. He strives to work with nature to produce sustainable, ecological, and beautifully designed gardens that

support local wildlife and promote biodiversity.

DUANE MCKENNA is a Distinguished University Professor and the Hill Professor of Biology in the Department of Biological Sciences at the University of Memphis (US), where he is founding director of the Center for Biodiversity Research. He studies insect systematics, genomics, evolution, and biodiversity, with a focus on beetles and insect–plant interactions.

LEIGH MCMAHON has a keen interest in natural history, in particular entomology. She is volunteering with the Diptera department at the Natural History Museum while completing a Master's degree in museum studies. Leigh is a keen naturalist, and her love of entomology developed when working as a museum curator in Paraguay, where her research interests included Lepidoptera.

SARAH MEREDITH is Conservation Project Officer for the Royal Entomological Society, undertaking scientific insect research and landscape scale conservation. She has worked on many butterfly and ant projects, including reintroducing the large blue to the UK. Sarah studied for an MSc in Wildlife Management and Conservation at University of Reading, focusing on the small heath.

AMPARO MORA has lived and worked as a biologist at Picos de Europa Nacional Park (Spain) since 2002. She is a mother of two and is currently finishing her PhD on butterfly ecology. Amparo loves books and nature.

GAIA MORTIER, currently a postgraduate researcher at the University of Reading, is dedicated to uncovering valuable information held by external parasites found within archaeological and museum collections. Driven by a passion for science communication, she advocates for the appreciation of the uncharismatic insect species that contribute to nature's intricate balance.

MUSONDA MOSES, a dedicated Zambian entomology PhD student at the University of Zambia, serves as a Trustee for the Royal Entomological Society. Passionate about malaria vectors and insect science, he contributes significantly to research and knowledge in these fields, aiming to make meaningful advancements in public health.

IMOGEN NEWENS-HILL was born in north London. She is an invertebrate keeper at Chester Zoo, specializing in the captive breeding of critically endangered species and *in situ* and *ex situ* conservation. Qualifications include Masters by Research in Entomology from the University of Reading and BSc in Zoology from the University of Sheffield.

YAMNI NIGAM is a Professor at Swansea University, leading the SU Maggot Research Group. Her team have investigated and published findings on molecules involved in larval therapy, secreted by the medicinal maggot *Lucilia sericata*. Yamni is currently leading the "Love a Maggot!" project, raising awareness and changing negative perceptions of maggots.

DARREN J. PARKER is an evolutionary biologist based at Bangor University. His work focuses on understanding how different reproductive strategies evolve in a wide variety of insects, including stick insects, flies, beetles, and crickets.

ROSE PEARSON is the librarian and archivist at the Royal Entomological Society. She graduated from University College London with an MA in library and information studies and has more than 10 years' experience working in libraries. She writes on rare books for *Antenna*, the Bulletin of the Royal Entomological Society.

TOM POPE is Reader in Entomology and Integrated Pest Management at Harper Adams University, a leading agricultural university in the UK. A Fellow of the Royal Entomological Society, Dr. Pope focuses his research on the development and implementation of sustainable approaches to crop protection.

MARK RAMSDEN is an entomologist specializing in the holistic management of invertebrates in agricultural environments. He has a background in applied biology and a PhD in methods for promoting beneficial insects around crops. Mark works with agricultural networks, demonstrating Holistic Integrated Pest Management and improving insect monitoring and management decisions.

STUART REYNOLDS is Emeritus Professor of Biology at the University of Bath and an Honorary Fellow and

former president of the Royal Entomological Society. He is interested in everything to do with insects, but especially what has made them the most evolutionarily successful of all animal groups.

JOE ROBERTS is an applied entomologist and lecturer in Integrated Pest Management at Harper Adams University. He studies plant–insect interactions to develop novel pest management tools that reduce societal reliance on synthetic chemical insecticides. Dr. Roberts is also a Fellow and Trustee of the Royal Entomology Society.

ELVA J. H. ROBINSON is a Professor in Ecology at the University of York and studies the organization of social animal groups, using ants as a model system. She researches ant behavior, evolution, and ecology to understand how these amazing creatures interact with each other and their environment.

CHARLIE ROSE is an avid naturalist with a long-standing fascination for insects. He has spent his career to date pursuing this passion, engaging with members of the public and private sectors on topics such as invertebrate ecology and population dynamics, as well as land management for the benefit of invertebrate communities.

MICHAEL J. SAMWAYS is Emeritus Distinguished Professor at Stellenbosch University, South Africa, specializing in insect conservation. He is a recipient of the Marsh Award from the Royal Entomological Society, the John Herschel Medal from the Royal Society of South

Africa, and the Gold Medal from the Academy of Science of South Africa.

MANU E. SAUNDERS is an ecologist and science communicator. Her main research focus is on insect communities and ecosystem function, particularly the important role that insects play in connecting humans and nature.

FRANCISCA SCONCE shares her passion for insects and entomology with the public. With a grounding in entomological research, she works in outreach and learning at the Royal Entomological Society. She has organized engagement activities since 2012 and works with audiences from a range of backgrounds.

LUIGI SEDDA (he/him) leads the Lancaster Ecology and Epidemiology Group, Lancaster University. His research focuses on spatial analyses of vector-borne diseases in humans and animals, primarily transmitted by insects. His work includes surveillance and quantifying health risks associated with the airborne spread of disease vectors, contributing to vital insights in the field.

DAVID SIMCOX has undertaken scientific work that has helped underpin the successful reintroduction of the large blue butterfly to the UK. This has been achieved in collaboration with a broad partnership of talented conservationists and with the dedication of his long-term colleagues, Jeremy Thomas and Sarah Meredith.

BRENT SINCLAIR grew up in New Zealand, did his PhD at the University of Otago, and postdocs in South Africa and Las Vegas. He studies the physiology and ecology of insect thermal biology and overwintering. He is currently a Professor of Biology at the University of Western Ontario, Canada.

SANDY SMITH specializes in forest entomology, forest health, and the ecology and biological management of invasive forest species at the Institute of Forestry & Conservation, University of Toronto. She has supervised over 65 graduate theses, published 160+ articles, taught over 30 university courses, and served on numerous national professional scientific and conservation organizations.

PETER SMITHERS is passionate about the way that insects are perceived. To engage the public, he organizes exhibitions and conferences that explore the relationship between entomology and art, but he has also flirted with entomologically inspired performance in the form of contemporary dance and opera. He continues to explore possibilities.

JEROEN VAN STEENIS is a Dutch self-learned entomologist who has been interested in Syrphidae since he was 14 years old and a member of the Dutch Youth Nature organization. His focus shifted from exclusive fieldwork to a wider perspective, including publishing scientific papers. His latest achievement was the founding of the Syrphidae Foundation and the *Journaal van Syrphidae*.

NIGEL STORK is a tropical entomologist and ecologist and Emeritus Professor. He was a pioneer of forest canopy studies. After studying at the University of Manchester, he worked at London's Natural History Museum. He moved to Australia in 1995 to lead the national rainforest research center, retiring from Griffith University in 2016.

JANE STOUT is Professor of Ecology at Trinity College Dublin. She is a Fellow of the Royal Entomological Society, an expert in pollinator and pollination ecology, and a prominent voice for biodiversity and its value. Jane works across disciplines, with a range of stakeholders, to improve environmental policy and practice.

SEIRIAN SUMNER is Professor of Behavioural Ecology at University College London. She studies social insects but is especially fond of wasps. She is co-founder of the citizen science project, Big Wasp Survey, and author of the popular science book *Endless Forms: Why You Should Love Wasps*.

JEREMY THOMAS spent 50 years researching the ecology and conservation of European butterflies and their relationships with ants. He was director of the Natural Environment Research Council's CEH Dorset Laboratory before becoming Professor of Ecology at the University of Oxford. He is a former president of the Royal Entomological Society.

DANA VANLANDINGHAM is a Professor at the College of Veterinary Medicine at Kansas State University, where she conducts research on mosquitoes and the viruses they transmit to both animals and humans. She has published over 100 peer-reviewed articles and has received numerous national and international awards for her work.

IAN WALLACE did his PhD on caddis larvae at Newcastle University and kept up that interest while a general curator of invertebrates at World Museum Liverpool, where he worked for 40 years. He has a passion for recording caddis distribution and writing keys to help others do so.

EMMA N. I. WEEKS is an entomologist whose research and writing focuses on medical and veterinary entomology, particularly arthropod behavior, chemical ecology, biological control, and integrated pest management. She currently is a self-employed editor based in northern Spain and enjoys reading, hiking, and observing nature.

ASHLEIGH WHIFFIN is an entomologist and museum curator. Her early research was in forensic entomology, which led to her specializing in carrion beetles. She is passionate about promoting these unappreciated insects and has helped establish a national recording scheme for the group.

JIM WHITFIELD is Professor Emeritus at the University of Illinois and co-author of the textbook *Introduction to Insect Biology and Diversity*. He specializes in systematics and evolution of small wasps that parasitize caterpillars and carry symbiotic viruses. He currently lives on the south coast of Cornwall.

ANDREW WHITTINGTON developed a childhood love for insects into an insect identification business, after graduating with a degree in Botany and Entomology and first working as a museum curator. He has traveled widely doing fieldwork and has published numerous papers and book chapters concerning insects.

CHARLIE WOODROW is an early career researcher in entomology and biophysics. He completed his PhD at the University of Lincoln, studying the evolution of sound production and hearing in Orthoptera. Beyond this, Charlie has ongoing projects in insect jumping, bee pollination biomechanics, and paleontology.

ROGER S. WOTTON is Emeritus Professor of Biology at UCL, where he taught Aquatic Biology using an integrated approach to the study of water bodies, from streams to oceans. Since retirement, he has developed his interest in art history and Victorian natural history and writes and lectures in both areas.

CHARLOTTE WRIGHT is a PhD student in the Tree of Life program at the Wellcome Sanger Institute. She uses genomic approaches to study the evolution of butterflies and moths to deepen our understanding of the genetic basis of diversification in this group.

EDITORS

EMILIE AIMÉ is Director of Publishing at the Royal Entomological Society, where she oversees the strategic direction of the society's books, journals, and membership magazine. She has worked in academic publishing for a number of societies and commercial publishers, mainly on peer-reviewed journals, for around 15 years. Her academic background is in zoology, and she is a lifelong nature lover.

JANE K. HILL is an ecologist at the University of York (UK). She is a trustee and president of the Royal Entomological Society (2022–2024), and she has been carrying out ecological research on insects for more than 30 years. Jane's research examines how insects are responding and adapting to climate change, and it has demonstrated how species are moving uphill and northward (in Europe) to track climate. She researches how habitat fragmentation and degradation is affecting these range shifts, and the benefits of Protected Areas and improving connectivity for conserving and enhancing insect biodiversity.

HELEN ROY is an ecologist with the UK Centre for Ecology & Hydrology and University of Exeter. Helen's research focuses on the ways in which environmental change affects biodiversity, particularly insects. She enjoys sharing her enthusiasm and engaging people with entomology through community science and co-leads the UK Ladybird Survey. In recent years, her research on biological invasions has increasingly been used to inform policy, and she co-chaired the Intergovernmental Science-Policy Platform on Biodiversity and Ecosystem Services assessment on invasive alien species and their control. Helen is an Honorary Fellow and past president of the Royal Entomological Society.

ALLAN WATT has conducted research on a range of insects, from the pine beauty moth in the north of Scotland to ants in the forests of Cameroon. After working on pests of agriculture (particularly cereal aphids) and forestry, he shifted focus to work on the factors affecting biodiversity in tropical forest and European landscapes. Most of his research has been done while working for the UK Centre for Ecology & Hydrology, and he currently focuses on building research capacity on biodiversity in Europe through the Alternet network. Involvement in the Royal Entomological Society has included editing *Agricultural and Forest Entomology* and *Antenna*.

ACKNOWLEDGMENTS

THE ROYAL ENTOMOLOGICAL SOCIETY

The editors of this book would like to thank the many people involved in bringing this unique collaborative project to life. First and foremost, we thank all the fantastic authors. Following an open call to the membership of the Royal Entomological Society, we received responses from 263 volunteers willing to be contributors. We were sorry not to involve everyone, but we were heartened and encouraged by this incredible response from our community. However, there were more than 90 contributors to this book (pp.249–254), from academic researchers and dedicated amateurs to industry specialists, with expertise spanning the wide discipline of entomology. Each topic was written by an insect expert, drawn from this inspiring pool of volunteers from the membership of the Royal Entomological Society. Putting together all of their contributions into an exciting and coherent book has been a challenge and hugely enjoyable.

We would like to offer particular thanks to Jim Hardie and Richard Harrington for their rapid responses to our many and varied queries and their exceptional breadth of knowledge of entomology. We would also like to thank Beulah Garner and Shaun Winterton for their fantastic taxonomic expertise. We have received incredible editorial support from Jen Banfield Zanin at the Royal Entomological Society, and her usual thoughtful efficiency, including in dealing with proofs among many other tasks, was much appreciated by everyone involved.

The amazing team at DK and their external contributors guided us throughout the process. Barbara Zuniga and Vicky Read, the design of the book is beautiful, and Peter Bull, Dan Crisp, and Stuart Jackson-Carter, the illustrations are incredible. Thank you for your patience in addressing all of our suggestions! Becky Gee, thank you for doing such a great job editing the text—we know it was not straightforward at times, but we appreciated your constructive comments that gave us much to think about! Finally, thanks to Sophie Blackman and Lucy Sienkowska for managing the project so expertly and for being such a pleasure to work with. We hope you have enjoyed immersing yourselves in the wonders of the amazing world of insects!

DORLING KINDERSLEY

The publisher would like to thank Aditya Kaytal for clearing image permissions, Kathy Steer for proofreading, and Ruth Ellis for providing the index.

DK | Penguin Random House

Senior Editors Sophie Blackman, Lucy Sienkowska
Senior US Editor Kayla Dugger
Executive US Editor Lori Cates Hand
Senior Designer Barbara Zuniga
Managing Editor Ruth O'Rourke
Senior Production Editor Tony Phipps
Production Controller Samantha Cross
DTP and Design Coordinator Heather Blagden
Sales Material & Jackets Coordinator Emily Cannings
Jacket Designer Izzy Poulson
Art Director Maxine Pedliham
Publishing Director Katie Cowan

Editorial Becky Gee
Design Vicky Read
Illustration Peter Bull Art Studio, Dan Crisp,
Stuart Jackson-Carter
Design styling concept Giulia Garbin

First American Edition, 2024
Published in the United States by DK Publishing,
a division of Penguin Random House LLC
1745 Broadway, 20th Floor, New York, NY 10019

Copyright © 2024 Dorling Kindersley Limited
24 25 26 27 28 10 9 8 7 6 5 4 3 2 1
001–342100–Oct/2024

A catalog record for this book
is available from the Library of Congress.
ISBN 978-0-5938-4349-9

Printed and bound in Slovakia
www.dk.com

MIX
Paper | Supporting
responsible forestry
FSC™ C018179

This book was made with Forest
Stewardship Council™ certified
paper—one small step in DK's
commitment to a sustainable future.
Learn more at **www.dk.com/uk/**
information/sustainability

PICTURE CREDITS

The publisher would like to thank the following for their kind
permission to reproduce their illustrations and photographs:

(Key: a-above; b-below/bottom; c-center; f-far; l-left; r-right; t-top)

Alamy Stock Photo: Impaint 228 bc; The Protected Art
Archive 228 cla; **Biodiversity Heritage Library:**
Metamorphosis insectorum surinamensium, Amsterdam:
Voor den auteur, als ook by G. Valck,[1705]. biodiversitylibrary.
org / page / 4139874 227; **Bridgeman Images:** © Look and Learn
231 bl; **Dan Crisp:** 1, 2, 5, 6, 7, 8, 11, 15, 17, 19, 22, 24, 35, 36, 39,
40, 45, 49, 51, 52-53, 54, 59, 60, 63, 71, 73, 74, 77, 80, 85, 86, 89,
94, 100, 104, 105, 106, 109, 111, 113, 114, 117, 118-119, 120, 123,
125, 128-129, 131, 133, 135, 138-139, 140, 143, 178, 181, 187, 192,
196, 200, 203, 205, 206, 209, 212, 222, 233; **Peter Bull Art
Studio:** 12, 13, 89, 123, 127, 144–177; **PLoS Biology:** Illustration
adapted from- © 2022 Cheng, Mishra, Cheng L., Mishra H.
(2022) Why did only one genus of insects, Halobates, take to
the high seas? PLoS Biol 20(4): e3001570. https://doi.org/
10.1371/journal.pbio.3001570, CC BY 4.0 DEED; Attribution 4.0
International- https://creativecommons.org/licenses/by/4.0/ 127
b; **Stuart Jackson-Carter:** 21, 22, 26, 27, 29, 31, 33, 42, 45, 47,
54, 57, 65, 67, 69, 79, 82, 91, 93, 97, 109, 127, 137, 189, 199, 215,
219, 221, 224.

All other images © Dorling Kindersley Limited

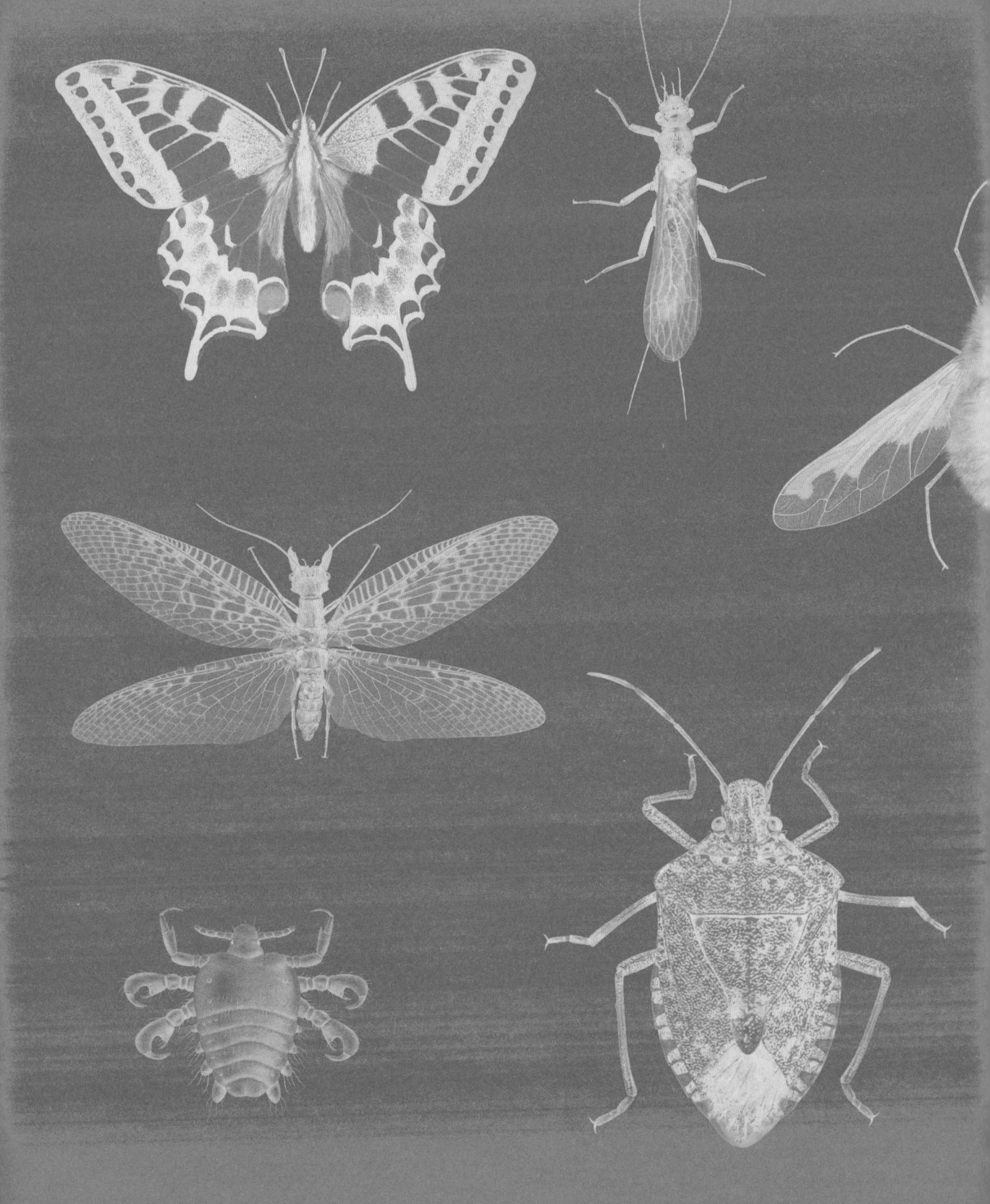